Review of VA Clinical Guidance for the Health Conditions Identified by the Camp Lejeune Legislation

Committee on the Review of Clinical Guidance for the Care of Health Conditions Identified by the Camp Lejeune Legislation

Board on the Health of Select Populations

INSTITUTE OF MEDICINE
OF THE NATIONAL ACADEMIES

THE NATIONAL ACADEMIES PRESS
Washington, D.C.
www.nap.edu

THE NATIONAL ACADEMIES PRESS 500 Fifth Street, NW Washington, DC 20001

NOTICE: The project that is the subject of this report was approved by the Governing Board of the National Research Council, whose members are drawn from the councils of the National Academy of Sciences, the National Academy of Engineering, and the Institute of Medicine. The members of the committee responsible for the report were chosen for their special competences and with regard for appropriate balance.

This study was supported by Contract/Grant No. VA241-P-2024 between the National Academy of Sciences and the Department of Veterans Affairs. Any opinions, findings, conclusions, or recommendations expressed in this publication are those of the author(s) and do not necessarily reflect the views of the organizations or agencies that provided support for the project.

International Standard Book Number-13: 978-0-309-31600-2
International Standard Book Number-10: 978-0-309-31600-6

Additional copies of this report are available for sale from the National Academies Press, 500 Fifth Street, NW, Keck 360, Washington, DC 20001; (800) 624-6242 or (202) 334-3313; http://www.nap.edu.

For more information about the Institute of Medicine, visit the IOM home page at: **www.iom.edu.**

Printed in the United States of America

The serpent has been a symbol of long life, healing, and knowledge among almost all cultures and religions since the beginning of recorded history. The serpent adopted as a logotype by the Institute of Medicine is a relief carving from ancient Greece, now held by the Staatliche Museen in Berlin.

Suggested citation: IOM (Institute of Medicine). 2015. *Review of VA clinical guidance for the health conditions identified by the Camp Lejeune legislation.* Washington, DC: The National Academies Press.

"Knowing is not enough; we must apply.
Willing is not enough; we must do."
—Goethe

INSTITUTE OF MEDICINE
OF THE NATIONAL ACADEMIES

Advising the Nation. Improving Health.

THE NATIONAL ACADEMIES

Advisers to the Nation on Science, Engineering, and Medicine

The **National Academy of Sciences** is a private, nonprofit, self-perpetuating society of distinguished scholars engaged in scientific and engineering research, dedicated to the furtherance of science and technology and to their use for the general welfare. Upon the authority of the charter granted to it by the Congress in 1863, the Academy has a mandate that requires it to advise the federal government on scientific and technical matters. Dr. Ralph J. Cicerone is president of the National Academy of Sciences.

The **National Academy of Engineering** was established in 1964, under the charter of the National Academy of Sciences, as a parallel organization of outstanding engineers. It is autonomous in its administration and in the selection of its members, sharing with the National Academy of Sciences the responsibility for advising the federal government. The National Academy of Engineering also sponsors engineering programs aimed at meeting national needs, encourages education and research, and recognizes the superior achievements of engineers. Dr. C. D. Mote, Jr., is president of the National Academy of Engineering.

The **Institute of Medicine** was established in 1970 by the National Academy of Sciences to secure the services of eminent members of appropriate professions in the examination of policy matters pertaining to the health of the public. The Institute acts under the responsibility given to the National Academy of Sciences by its congressional charter to be an adviser to the federal government and, upon its own initiative, to identify issues of medical care, research, and education. Dr. Victor J. Dzau is president of the Institute of Medicine.

The **National Research Council** was organized by the National Academy of Sciences in 1916 to associate the broad community of science and technology with the Academy's purposes of furthering knowledge and advising the federal government. Functioning in accordance with general policies determined by the Academy, the Council has become the principal operating agency of both the National Academy of Sciences and the National Academy of Engineering in providing services to the government, the public, and the scientific and engineering communities. The Council is administered jointly by both Academies and the Institute of Medicine. Dr. Ralph J. Cicerone and Dr. C. D. Mote, Jr., are chair and vice chair, respectively, of the National Research Council.

www.nationalacademies.org

COMMITTEE ON THE REVIEW OF CLINICAL GUIDANCE FOR THE CARE OF HEALTH CONDITIONS IDENTIFIED BY THE CAMP LEJEUNE LEGISLATION

DAVID R. NERENZ (*Chair*), Director of the Center for Health Policy and Health Services Research, Henry Ford Health System, Detroit, MI
ROBERT J. ALPERN, Dean, Yale Medical School, New Haven, CT
PAOLO BOFFETTA, Director of the Institute for Translational Epidemiology and Associate Director for Population Sciences, The Tisch Cancer Institute, Icahn School of Medicine at Mount Sinai Hospital, New York
MARY E. DAVIS, Professor, West Virginia University Health Sciences Center, Morgantown, WV
MICHAEL E. GOLDBERG, David Mahoney Professor of Brain and Behavior, Columbia University College of Physicians and Surgeons, New York
PAUL GRUNDY, Director of Healthcare Transformation, IBM, Hopewell Junction, NY
NANCY L. KEATING, Professor of Health Care Policy, Harvard Medical School, Boston, MA
PATRICIA JANULEWICZ LLOYD, Assistant Professor, Boston University School of Public Health, Boston, MA
GARY O. RANKIN, Professor and Chair, Joan C. Edwards School of Medicine, Marshall University, Huntington, WV
MARK J. UTELL, Professor, University of Rochester School of Medicine, Rochester, NY
CAROL S. WOOD, Oak Ridge National Laboratory, Oak Ridge, TN
ALBERT W. WU, Professor and Director, Center for Health Services and Outcomes Research, Johns Hopkins Bloomberg School of Public Health, Baltimore, MD

IOM Staff

ROBERTA WEDGE, Study Director
CARY HAVER, Associate Program Officer
HEATHER CHIARELLO, Research Associate
ADRIANNA MOYA, Senior Program Assistant (through May 2014)
FREDRICK ERDTMANN, Director, Board on the Health of Select Populations

Reviewers

This report has been reviewed in draft form by individuals chosen for their diverse perspectives and technical expertise, in accordance with procedures approved by the National Research Council's Report Review Committee. The purpose of this independent review is to provide candid and critical comments that will assist the institution in making its published report as sound as possible and to ensure that the report meets institutional standards for objectivity, evidence, and responsiveness to the study charge. The review comments and draft manuscript remain confidential to protect the integrity of the deliberative process. We wish to thank the following individuals for their review of this report:

Peter F. Buckley, Georgia Regents University
Matthew Cave, University of Louisville
Evan D. Kharasch, Washington University in St. Louis
Francine Laden, Harvard School of Public Health
Bruce P. Lanphear, Simon Fraser University
Lawrence H. Lash, Wayne State University School of Medicine
Bruce S. McEwen, The Rockefeller University
Martin A. Philbert, University of Michigan

Although the reviewers listed above have provided many constructive comments and suggestions, they were not asked to endorse the conclusions or recommendations nor did they see the final draft of the report before its release. The review of this report was overseen by **Lynn R. Goldman,** George Washington University School of Public Health and Health Sciences, and **Kenneth W. Kizer,** University of California, Davis, School of Medicine. Appointed by the National Research Council and the Institute of Medicine, they were responsible for making certain that an independent examination of this report was carried out in accordance with institutional procedures and that all review comments were carefully considered. Responsibility for the final content of this report rests entirely with the authoring committee and the institution.

Contents

Acronyms and Abbreviations

ADHD	attention deficit hyperactivity disorder
AFLD	alcohol-related fatty liver disease
ALS	amyotrophic lateral sclerosis
ATSDR	Agency for Toxic Substances and Disease Registry
CI	confidence interval
CKD	chronic kidney disease
CNS	central nervous system
CT	computerized tomography
CYP3A	cytochrome P4503A
DCE	dichloroethylene
DCVC	dichlorovinylcysteine
DCVCS	dichlorovinylcysteine sulfoxide
DCVG	dichlorovinyl glutathione
DCVT	dichlorovinylthiol
eGFR	estimated glomerular filtration rate
EPA	U.S. Environmental Protection Agency
ESRD	end-stage renal disease
FMO3	flavin monooxygenase 3
GAO	Government Accountability Office
GST	glutathione S transferase
HIV	human immunodeficiency virus
HR	hazard ratio

IARC	International Agency for Research on Cancer
IOM	Institute of Medicine
IRIS	Integrated Risk Information System (of the EPA)
KIM-1	kidney injury molecule-1
MCL	maximum contaminant level
MRI	magnetic resonance imaging
MS	multiple sclerosis
NAcDCVC	N-acetyl dichlorovinylcysteine
NAcDCVCS	N-acetyl dichlorovinylcysteine sulfoxide
NAcTCVC	N-acetyl trichlorovinylcysteine
NAcTCVCS	N-acetyl trichlorovinylcysteine sulfoxide
NAFLD	non-alcoholic fatty liver disease
NAG	N-acetyl-β-D-glucosaminidase
NAT	N-acetyltransferase
NIOSH	National Institute for Occupational Safety and Health
NMDA	N-methyl-D-aspartic acid
NRC	National Research Council
NTP	National Toxicology Program
OCD	obsessive compulsive disorder
OR	odds ratio
OSHA	U.S. Occupational Safety and Health Administration
PCE	perchloroethylene (or tetrachloroethylene)
Pi-GST	Pi-glutathione S transferase
PTSD	posttraumatic stress disorder
RR	risk ratio
SIR	standardized incidence ratio
SMR	standardized mortality ratio
TAFLD	toxicant-associated fatty liver disease
TCA	trichloroacetic acid
TCE	trichloroethylene
TCOH	trichloroethanol
TCVC	trichlorovinylcysteine
VA	U.S. Department of Veterans Affairs
VHA	Veterans Health Administration (VA's health care system)
γGTP	γ-glutamyltranspeptidase

Summary

U.S. Marine Corps Base Camp Lejeune covers about 156,000 acres in eastern North Carolina, and at any given time is home to about 170,000 active-duty personnel, family members, retirees, and civilian employees who live on base or in the surrounding community. Between 1957 and 1987, the groundwater at Camp Lejeune was inadvertently contaminated with chemicals, primarily industrial solvents. Many of these chemicals were later found to cause cancer and other health problems, although not all of them were recognized as toxic at the time of contamination. In 1980 trichloroethylene (TCE) and perchloroethylene (PCE, also called tetrachloroethylene), as well as other solvents, were first detected at Camp Lejeune in treated drinking water, and by 1987 the contaminated water wells were closed. In 1989, the U.S. Environmental Protection Agency (EPA) placed Camp Lejeune on the National Priorities List, also known as Superfund. It is estimated that between 500,000 and 1 million people may have used the contaminated water, and many of them continue to have concerns about the long-term health effects that might result from that exposure.

STUDIES ON THE CAMP LEJEUNE POPULATION

From 1991 to 1997 the Agency for Toxic Substances and Disease Registry (ATSDR), part of the Centers for Disease Control and Prevention, conducted a public health assessment that evaluated exposures and potential risks at Camp Lejeune. It also performed a historical reconstruction of the contamination based on water quality modeling and estimated that well contamination with PCE from an off-base dry cleaner began as early as the 1950s.

In 2009, the National Research Council (NRC) released its report *Contaminated Water Supplies at Camp Lejeune: Assessing Potential Health Effects* in response to a request from Congress. That report built on a 2003 Institute of Medicine (IOM) report that reviewed the toxicologic and epidemiologic literature on solvents that had been used in the 1990–1991 Gulf War and their potential health effects. The NRC report assessed studies published after the IOM report and focused on the potential health effects of the solvents on Camp Lejeune residents and similarly exposed populations.

In 2012, Congress passed the Honoring America's Veterans and Caring for Camp Lejeune Families Act, also known as the Janey Ensminger Act (P.L. 112-154). The act provides health benefits to veterans and family members who have any of 15 health conditions.[1] Eligible veterans must have served on active duty at Camp Lejeune for 30 days or longer between January 1, 1957, and December 31, 1987, and eligible family members must have resided (including being in utero to a mother in residence) at Camp Lejeune for 30 days or longer during the same time frame.

To assist in the implementation of the act, the Veterans Health Administration (VHA) has drafted clinical guidance, including five clinical algorithms, to help health care providers determine whether a veteran or family member has a medical condition that is covered by the act and whether an episode of care is related to a covered condition.

COMMITTEE'S STATEMENT OF TASK AND APPROACH

To ensure that the clinical guidance for the 15 covered medical conditions listed in Public Law 112-154 is "scientifically sound," the U.S. Department of Veterans Affairs (VA) asked the IOM to convene an ad hoc committee to review the guidance for VHA staff and make recommendations for its improvement. In addition, the committee was asked to perform the following specific tasks:

1. Based on the latest scientific literature and the committee's review, describe the medical conditions that result from "renal toxicity" due to solvent exposures.
2. Based on the latest scientific literature and the committee's review, characterize the "neurobehavioral effects" as mandated for coverage in the law.

To conduct its task, the committee held two open sessions to learn about the guidance from VA health professionals. Literature searches were also performed to identify the recent epidemiologic and toxicologic studies and assessments of the contaminants of interest. The committee reviewed the available literature to identify possible renal and neurobehavioral endpoints; no endpoints were ruled out by the committee in advance.

The committee adopted a rule that a renal or neurobehavioral effect must be reported with statistical significance in at least one relatively well-designed study, or have sufficient strength of evidence to be considered a possible effect. In cases where the weight of the evidence was sparse but showed a positive association, or was equivocal and expert judgment was used to make the finding, the committee gave the benefit of the doubt to the veteran and family member.

The committee also recognized that many factors can affect the etiology and presentation of a health condition. This is of particular concern because many of the health conditions in the VA clinical guidance require that the clinician determine whether the health condition was caused by something other than the patient's residence at Camp Lejeune during the time of contamination. VA has made a policy decision that clinicians do not have to consider exclusionary factors for any of the cancers or for scleroderma, but these factors are considered for the other health conditions. Because these are policy decisions and not based on scientific evidence, the committee did not comment on the validity of these decisions.

RENAL TOXICITY

Toxicologic and Epidemiologic Evidence

Previous reviews have suggested that among the contaminants residents at Camp Lejeune were exposed to TCE and PCE were the most likely to be responsible for acute kidney injury and subsequent chronic renal disease. In general, human and animal studies demonstrate that high-dose exposures are required for acute renal effects to be observed and that such effects are variable among species. In animal studies, acute exposure to high doses of

[1] The 15 conditions in the act are esophageal cancer, lung cancer, breast cancer, bladder cancer, adult leukemia, kidney cancer, multiple myeloma, myelodysplastic syndromes, female infertility, miscarriage, hepatic steatosis, scleroderma, renal toxicity, and neurobehavioral effects identified by the 2009 NRC report as having limited/suggestive evidence of an association with exposure to TCE, PCE, or solvent mixtures; and non-Hodgkin's lymphoma.

TCE causes tubular necrosis, which results in a decreased glomerular filtration rate. In both humans and animals, chronic exposures to high doses of either TCE or PCE cause kidney pathology, such as cytomegaly, karyomegaly, and necrosis of the tubular epithelium.

Based on the cumulative data, there appears to be strong evidence for an association between acute exposure to high levels of TCE or PCE and acute tubular toxicity in both rodents and humans, although humans metabolize these chemicals to a lesser extent and are thus more resistant to their adverse effects. There is accumulating evidence that acute renal injury, as might occur soon after exposure, significantly increases the likelihood that chronic kidney disease will appear many years later; such an effect can occur even if the acute injury is subclinical and thus not detected at the time of exposure. Thus, a patient should not be ineligible for the VA program because of a lack of documented evidence of kidney disease during or shortly after residence at Camp Lejeune.

While there is some evidence for increased mortality from solvent-induced hypertensive end-stage renal disease (ESRD), the association between TCE and PCE and chronic kidney disease is less clear, although there does appear to be an association between exposures to high levels of these solvents and ESRD. However, the documented levels of PCE and TCE in the drinking water at Camp Lejeune were much lower than those in the human and animal studies reviewed in this report, and the exposure duration would likely have been much shorter for Camp Lejeune residents. There is no evidence for an increased incidence of chronic kidney disease in those who resided at Camp Lejeune during the time of the contaminated drinking water. Nevertheless, although the evidence indicates that chronic kidney disease in Camp Lejeune residents is likely due to other causes, the role for solvent exposure cannot entirely be ruled out. This is a common problem when seeking causes of kidney disease where there is no specific diagnostic histopathology.

Clinical Guidance and Algorithm for Renal Toxicity

The VA guidance asks first whether the patient has evidence of renal injury, when the onset of chronic kidney disease occurred, and if the patient has other comorbid conditions. The clinician then assesses whether it is probable that the chronic kidney disease is attributable to a known cause other than solvent toxicity. If there is no evidence for another cause, chronic kidney disease could be due to toxic exposure. The committee finds that VA's general approach to renal toxicity in the guidance and in algorithm K is appropriate and in cases of uncertainty with regard to the etiology of the renal injury, the case should be resolved in favor of the veteran or family member.

There may be a lack of evidence of acute renal nephrotoxicity at the time of exposure because it did not occur, because a patient was asymptomatic and there was no indication to conduct the necessary laboratory tests, or because the tests were not sensitive enough to detect mild disease. Neither the guidance nor the algorithm includes other indicators of acute renal injury, such as abnormal urinalysis results, serum creatinine, or blood urea nitrogen which, if assessed at about the time of exposure and documented in medical records, may help a clinician establish that acute effects had occurred, which later might result in or contribute to chronic kidney disease. The committee finds that these types of tests, conducted while the patient was in residence at Camp Lejeune, should be considered when determining whether or not the patient's chronic kidney disease is related to exposure to contaminated drinking water while at Camp Lejeune. If the evaluation shows that the patient's kidney disease is compatible with another etiology, such as diabetic nephropathy or hypertensive nephrosclerosis, it is unlikely that solvent exposure at Camp Lejeune was the causative agent. If the evaluation does not suggest another etiology, or if the clinical course is atypical for the identified etiology, the patient should be included in the Camp Lejeune program.

> **Therefore, the committee recommends that VA consider modifying the guidance and algorithm K—as suggested in revised algorithm K—to indicate that patients presenting with defined reductions in glomerular filtration rate or proteinuria AND who had abnormal renal function tests or urinalysis of unknown etiology while residing at Camp Lejeune should be accepted to the program. The committee also recommends that VA consider accepting into the Camp Lejeune program patients with chronic kidney disease, but without evidence of kidney damage during or around the time of residence at Camp Lejeune, if there are no other more likely causes of their kidney disease.**

NEUROBEHAVIORAL EFFECTS

All Age Exposures

The 2009 NRC report found that there was limited/suggestive evidence of an association between solvent exposure and neurobehavioral effects including abnormal results on neurobehavioral test batteries; symptoms such as fatigue, lack of coordination, sensory disturbances, confusion, depression, tension, trouble concentrating, and headache; deficits in attention, reaction time, visuomotor coordination, motor function, digit symbol, and contrast sensitivity; and certain neuropsychological disorders such as learning or behavioral disorders. The 2009 NRC report found that most of the neurobehavioral effects in the epidemiologic studies were concurrent with exposure and that few studies assessed long-term effects after the exposure ended. That report further concluded that there is inadequate/insufficient evidence to determine whether an association exists between an exposure to solvents and amyotrophic lateral sclerosis (ALS), multiple sclerosis, Alzheimer's disease, or Parkinson's disease.

The NRC report separated neurologic diseases, such as Alzheimer's disease and Parkinson's disease, from neurobehavioral effects, which left unanswered what those signs and symptoms indicated in terms of diagnostic entities. Because neurobehavioral symptoms or testing can be indicative of neurologic or behavioral problems, in an effort to be complete, this committee chose to define neurobehavioral effects broadly to include all neurologic and behavioral effects (diseases, disorders, symptoms, and deficits). Toxicologic studies were not specifically described because the studies published since 2008 did not describe clinical outcomes that a physician might encounter in patients.

New studies assessed by this committee suggest that deficits in visuomotor function, motor function, and concentration (that is, attentional deficits) best characterize the long-term neurobehavioral symptoms and deficits associated with solvent exposure. No new evidence provided additional support for a relationship between an exposure to solvents and the development of ALS, multiple sclerosis, or Alzheimer's disease. However, the committee reviewed four new studies on Parkinson's disease and solvent exposure that led it to conclude that Parkinson's disease is a neurobehavioral effect that may result from exposure to TCE and/or PCE. Because of the slow onset of Parkinson's disease, patients who develop it years after their exposure, regardless of their age at Camp Lejeune, may not have had symptoms at the time of exposure.

The committee recommends that VA consider adding Parkinson's disease in the clinical guidance and in algorithm B as a neurobehavioral effect that may result from exposure to contaminated drinking water at Camp Lejeune.

In Utero and Childhood Exposures

At Camp Lejeune, pregnant women may have inadvertently exposed their fetuses to the contaminated water, and children may also have been exposed. The committee believes that the health impacts of contaminated Camp Lejeune water on fetuses, infants, and children needs to be considered in the VA guidance. The 2009 NRC report concluded "that there continues to be inadequate/insufficient evidence to determine whether an association exists between chronic exposure to TCE or PCE and congenital malformations" such as congenital heart defects, neural tube defects, and oral clefts. However, recent studies of congenital anomalies in children born to mothers exposed to TCE, PCE, or other solvents during pregnancy show a clear association with neural tube defects.

The committee recommends that VA consider adding neurobehavioral effects as a result of neural tube defects in the Camp Lejeune clinical guidance and in algorithm B-1.

Most of the new literature identified by the committee was the product of epidemiologic studies of a Cape Cod, Massachusetts, population that had been exposed to PCE in drinking water, from 1968 through 1980. The committee found that, in general, the Cape Cod community studies of PCE exposure had appropriate controls, appropriate adjustment for confounders, and large enough samples to allow for the detection of elevated risks.

They are the only studies that have examined psychological and psychosocial outcomes in association with in utero or childhood exposure to PCE, TCE, or other solvents. Thus, although the positive findings reported for the Cape Cod cohorts have not been confirmed by research in other populations, they have good scientific plausibility and a demonstrate a dose–response.

Committee members were not in agreement on whether the two studies on illicit drug use and bipolar disorder provided enough evidence to warrant a recommendation on the inclusion of these two neurobehavioral effects in the guidance and algorithms.

Nevertheless, in keeping with the VA policy that "in cases where there is reasonable doubt as to the diagnosis or primary cause for the diagnosis, clinicians should resolve in favor of the Camp Lejeune veteran or family member," the committee recommends that VA consider including adolescent and adult illicit drug use and bipolar disorder as neurobehavioral effects in the Camp Lejeune clinical guidance and algorithm B-1.

The committee acknowledges that the visual deficits found in the Cape Cod population and other studies are subclinical, that several studies had small sample sizes, and that evidence is lacking by which to assess whether the effects are short term or long term. Although prior reports found that there was inadequate/insufficient evidence to determine whether an association exists between exposure to solvents and long-term reduction in color discrimination, this committee finds that the weight of evidence indicates that deficits in contrast sensitivity and color discrimination may occur from such exposures.

The committee recommends that problems with contrast sensitivity and color discrimination be included in the clinical guidance and algorithm B as neurobehavioral effects that may result from exposure to contaminated drinking water at Camp Lejeune, although it recognizes that these are typically subclinical (that is, they are not detectable upon routine examination), and no treatments for them are currently available. Given their subclinical nature, the committee further recommends that patients not be screened for these conditions unless there is a clear reason to do so (for example, the patient reports visual problems), and that the results of any screening or testing for visual problems should be noted in the patient's record.

Revising the Guidance and Algorithm

The guidance currently has a short section for clinicians on what is meant by neurobehavioral effects, what would be a covered condition, and what signs or symptoms should be determined to have been present when a veteran or family member was exposed to contaminated drinking water during or shortly after residence at Camp Lejeune. It is unclear how VA selected the neurobehavioral effects given in the guidance and algorithm B, why the effects do not reflect the 2009 NRC report, or why neurotoxic endpoints from a 1987 National Institute for Occupational Safety and Health Current Intelligence Bulletin on occupational exposure to organic solvents are included. The committee notes that more recent reviews are available from organizations such as EPA.

The committee recommends that the VA clinical guidance and algorithm B be revised to be consistent and to reflect recent literature.

The guidance does not currently address conditions associated with in utero or childhood exposures at Camp Lejeune. The committee believes it is important that the guidance address these exposures because those outcomes differ from those for adults and are not captured in the current guidance or in algorithm B.

Thus, the committee recommends that VA consider including in the clinical guidance a new algorithm B-1 for neurobehavioral effects specific to prenatal and childhood exposure at Camp Lejeune.

OTHER HEALTH OUTCOMES

Cancer and Neoplastic Diagnoses

The guidance is unclear or fails to address four issues regarding how or what cancers are covered by the Camp Lejeune program.

Eight cancers (esophageal cancer, lung cancer, breast cancer, bladder cancer, kidney cancer, leukemia, multiple myeloma, and non-Hodgkin's lymphoma) and myelodysplastic syndromes are listed in the 2012 Janey Ensminger Act. VA has stated that following the precedent it set in response to Agent Orange exposures for Vietnam veterans it will cover listed cancers regardless of latency because this policy provides the benefit of the doubt to veterans and family members.

The committee recommends that VA clearly state in the guidance its policy decision to not consider the latency of cancers.

VA may wish to clarify whether it will cover secondary or recurrent/metastatic cancers if the first primary (which was one of the eight neoplasms covered) occurred before the exposure at Camp Lejeune.

The committee recommends that VA include in the Camp Lejeune program patients with second primary cancers (but not recurrent or metastatic cancers) whose primary cancer was one of the covered cancers, even if their first primary cancer was diagnosed before residence at Camp Lejeune.

The guidance and algorithm do not address whether precancerous lesions of the eight cancers and myelodysplastic syndromes (such as ductal carcinoma in situ, Barrett's esophagus, and monoclonal gammopathy of undetermined significance) are also covered. VA has indicated that it plans to cover precancerous lesions, and the committee finds this approach to be reasonable.

The committee recommends that VA clearly address precancerous lesions in the clinical guidance and in the core algorithm.

Finally, the guidance defines active treatment for cancer as surgery, chemotherapy, or radiation therapy, or some combination of the three, but it does not specifically include immunotherapy and hormone therapy.

The committee recommends that VA specifically include hormonal treatment and immunotherapy as part of the "active treatment" for cancer in the clinical guidance.

Scleroderma (Systemic Sclerosis)

Scleroderma is a rare autoimmune condition characterized by the presence of thickened, sclerotic skin lesions. The 2009 NRC report concluded that there was limited/suggestive evidence of an association between mixed solvent exposures and scleroderma. New reviews conducted by other authoritative entities have confirmed the association between scleroderma and TCE. In 2013, the American College of Rheumatology updated its diagnostic criteria for scleroderma.

Because scleroderma onset can occur at any time after exposure to a solvent, any exposed veterans and family members are eligible for health benefits and are accepted to the Camp Lejeune program regardless of when their disease was diagnosed. The committee finds that the guidance and algorithm for scleroderma are reasonable and appropriate.

The committee recommends that VA update the guidance in accordance with the 2013 American College of Rheumatology diagnostic criteria for scleroderma.

Miscarriage and Infertility

Miscarriage refers to a spontaneous abortion. Factors that may increase the risk of miscarriage include maternal hormone problems or infections, trauma, age greater than 45 years, smoking, drug use, excessive caffeine, and exposure to the solvents such as those found at Camp Lejeune. Infertility—that is, failure to achieve a clinical pregnancy after 12 months or more of regular unprotected sexual intercourse—may be caused by several factors that affect different aspects of female reproduction. One such factor is exposure to environmental contaminants such as TCE and PCE, which can also affect fertility by reducing fecundity and altering menstrual cycles.

There is no evidence for an increased risk of miscarriage remaining after an exposure to solvents has ended, but a miscarriage may have long-term psychologic and medical consequences, such as depression, anxiety, and posttraumatic stress disorder (PTSD), which themselves may persist and result in long-lasting psychological, social, and health changes. Infertility may also have an impact on quality of life and mental health. Some studies indicate, for instance, that infertility is associated with depression and loneliness.

VA guidance on female infertility and miscarriage specifies time of onset. Exposed veterans and family members who experienced or were diagnosed with these problems during their time at Camp Lejeune are eligible for health benefits if they require ongoing medical treatment. Later or current infertility or miscarriage in a woman who was a child, adolescent, or young adult while at Camp Lejeune is not covered.

Algorithm W requires documentation that infertility or miscarriage occurred during residence on Camp Lejeune. The committee notes that medical records from that time may not be available, so it is important that VA encourage informed clinical judgment to identify veterans or family members who have persistent health problems that may have resulted from miscarriage or infertility that occurred at the time of exposure. The committee finds the guidance and algorithm for miscarriage and infertility to be generally appropriate.

The committee recommends that throughout the guidance and algorithm VA refer to "physical and mental health conditions" related to prior infertility or miscarriage, rather than to "medical conditions," "medical problem," or "medical treatment."

Hepatic Steatosis

Hepatic steatosis, also referred to as fatty liver, is associated with a variety of conditions, including type 2 diabetes, obesity, alcohol use, hepatitis, hyperlipidemia, and other liver diseases; the use of some medications (e.g., chemotherapeutic agents, antibiotics); and exposure to organic chemicals such as TCE, PCE, and chloroform. Most patients who have hepatic steatosis have elevated liver enzyme levels, but they may be asymptomatic. A probable diagnosis can be established by ultrasound, computed tomography, and magnetic resonance imaging. Hepatic steatosis is generally benign, and the condition is reversible, but if it persists more severe pathologies such as fibrosis, cirrhosis, and liver cancer may develop.

Application of the guidance and algorithm H for hepatic steatosis is challenging because of the high prevalence of other potential causes of hepatic steatosis in the general population. Informed clinical judgment can help identify veterans or family members whose hepatic steatosis may have resulted from exposure to drinking water at Camp Lejeune on the basis of its persistence since residing at Camp Lejeune and the absence of other more likely causes.

The guidance states "[M]oreover if a patient's clinical course is atypical or progresses faster than expected, then exacerbation by TCE, PCE or other organic solvents from Camp Lejeune should be considered." However, there is no evidence that solvent exposure would result in an atypical presentation or rapid progression of hepatic steatosis at a later date. Chronic alcohol consumption of 16 g of alcohol or more per day is strongly associated with steatosis.

Based on the evidence, the committee recommends that VA delete the phrase "atypical or progresses faster than expected" in the clinical guidance. The committee further recommends that VA replace the term "alcohol abuse," listed among the other causes of hepatic steatosis in the clinical guidance and algorithm, with "alcohol use ≥ 20 g/day for women or ≥ 30 g/day for men."

Finally, there are several commonly used medications that are known to cause steatosis, including chemotherapeutic agents and corticosteroids.

The committee recommends that VA include "some medications" in the list of other causes in algorithm H and that examples of those medications be listed in the text of the clinical guidance.

USE OF THE GUIDANCE

The committee was asked to assess the scientific soundness of the guidance for the covered health outcomes. The committee was not asked to comment on the implementation of, administration of, training for, or evaluation of the Camp Lejeune Health Program itself. It is important to remember that the guidance is based not only on scientific evidence, but also on VA policies and congressional legislation as well.

The guidance specifies that VA will reimburse eligible family members for screenings related to the 15 covered conditions if clinically indicated or if recommended by the U.S. Preventative Services Task Force only if the outcome of that screening leads to the diagnosis of a covered condition. The diagnosis of a covered condition may require screening as well as a diagnostic evaluation at the discretion of the clinician, but the guidance does not indicate whether a diagnostic evaluation will be covered.

The committee recommends that VA revise the sentence on page 3 of the guidance to read "VA will reimburse eligible family members for screening and diagnostic evaluations that are clinically indicated or recommended by the U.S. Preventive Services Task Force, and that lead to a diagnosis of a covered condition."

Decision Points

The committee considered the usefulness of the three decision points to assess whether a "medical illness, injury or condition is eligible for coverage under the Camp Lejeune Program." It indicated where revisions to the guidance and the algorithms might improve clarity and consistency.

(1) Does the Camp Lejeune program participant have one or more of the covered conditions?

The committee considered three topics for this decision point: referrals, secondary conditions, and time of symptom onset and duration. The guidance does not indicate when referrals to specialists, such as psychiatrists or nephrologists, may be appropriate for the diagnosis of a covered condition.

The committee recommends that referrals to specialists should be made when clinically indicated to obtain a definitive diagnosis and that VA should have a standardized process for making such referrals.

The guidance states that VA has authority to reimburse family members for medical conditions that are secondary to a covered condition. However, the descriptions and algorithms for the covered conditions, other than for female infertility and miscarriage, do not acknowledge that secondary conditions and medical complications can result not only from the presence of the condition itself, but also from disease progression and from treatment for the condition.

The committee recommends that VA should consider adding the need to diagnose and treat secondary conditions to the descriptions or algorithms for the covered primary conditions.

In general, the committee notes that the time of onset and duration are not specified for every covered condition. For example, VA has made a policy decision that the time of onset matters for miscarriage and infertility, but that it is not a consideration for cancer. This variation in criteria among outcomes may result in confusion on the part of the Camp Lejeune veterans, family members, and clinicians.

The committee recommends that VA specify details for the same domains (such as the criteria for diagnosis, onset, duration, and other possible causes and exclusionary factors) for all conditions in order to ensure consistency, completeness, and clarity.

(2) Is there evidence that the condition occurred as a result of a cause other than residence at Camp Lejeune?

The guidance states, "[H]ospital care and medical services may not be furnished . . . for an illness or condition of a Camp Lejeune Veteran or family member that is found, in accordance with guidelines issued by the Under Secretary for Health, to have resulted from a cause other than the residence at Camp Lejeune." The act requires other causes be assessed only for family members and not for veterans. The committee finds the language in the guidance regarding the need to consider other causes of covered conditions for veterans or family members to be inconsistent and unclear.

The committee recommends that VA state whether veterans must meet the same criteria as family members regarding other possible causes for a condition.

The guidance uses several terms to assist the clinician in determining whether the condition had another cause, such as "are as likely as not," or "probable," but no criteria are given for making such judgments. This is particularly problematic as the guidance states "In cases where there is reasonable doubt as to the diagnosis or primary cause for the diagnosis, clinicians should resolve in favor of the Camp Lejeune Veteran or family member." The committee finds that the language in the guidance is inconsistent with regard to the likelihood that exposure at Camp Lejeune resulted in a covered condition.

The committee recommends that VA set one standard for the likelihood that a condition (with the exception of cancer and scleroderma) must be related to residence at Camp Lejeune. The committee also recommends that VA reword the decision point to read "Is there evidence that the condition is as likely as not to have occurred as a result of a cause other than residence at Camp Lejeune?" in order to more accurately reflect the rest of the guidance.

(3) Is the episode of care or treatment related to the covered condition?

The guidance asks clinicians to "verify" or "certify" information pertaining to whether a specific visit, treatment, or secondary condition is related to a covered condition. It is unclear what health care providers, including non-VA clinicians, must do in order to certify or verify that a treatment or service is related to a covered condition and what documentation must be submitted.

The committee recommends that VA include instructions to clinicians about how to record essential information regarding their patients' diagnoses and treatments for those conditions.

1

Introduction

U.S. Marine Corps Base Camp Lejeune, located in eastern North Carolina, covers 156,000 acres and stretches along 11 miles of beach. It has maintained combat-ready marine units for expedited deployment since 1941. About 170,000 active-duty personnel, family members, retirees, and civilian employees live on base or in the surrounding community. The supporting infrastructure on base includes businesses, schools, recreational facilities, and municipal services such as a base landfill and water treatment system (U.S. Marine Corps, undated). In 2007, an estimated 54,000 people lived and worked on base, with families living on base for an average of 2 years (GAO, 2007).

Between 1957 and 1987 the groundwater at Camp Lejeune was contaminated with industrial chemicals, primarily chlorinated solvents such as trichloroethylene (TCE) and perchloroethylene (PCE, also called tetrachloroethylene). Many of these chemicals were later found to cause cancer and other health problems, although not all of them were recognized as toxicants at the time of contamination. The 30-year period of contamination, the lack of records documenting residence at Camp Lejeune during that time, and the transient nature of military assignments and deployments, make it almost impossible to know who or how many people were exposed. However, it is estimated that between 500,000 and 1,000,000 people may have been exposed (Walters, 2014).

A BRIEF HISTORY OF CAMP LEJEUNE WATER CONTAMINATION

In the 1980s, Camp Lejeune obtained its drinking water from as many as eight water systems fed by more than 100 wells that pumped water from a freshwater aquifer approximately 180 feet below ground. Drinking water was made from treated groundwater supplied by a rotating combination of multiple wells so that not all wells were providing water to a system at any given time. It was in the 1980s that volatile organic compounds—including chlorinated solvents such as PCE and TCE and their degradation products, aromatic solvents such as benzene, and other organic compounds such as vinyl chloride—were detected at Camp Lejeune in two separate water systems—Hadnot Point and Tarawa Terrace—that served base housing areas. Tarawa Terrace was contaminated primarily with PCE and its degradation products; Hadnot Point's major contaminants were TCE and its degradation product trans-1,2-dichloroethylene (DCE). At that time no action was taken because there was little knowledge about the toxicity of TCE and PCE, there were no drinking water regulations with enforceable limits for these chemicals, and there was uncertainty about the validity of the water tests due to variations in results. Nevertheless, base officials removed 10 contaminated wells from service. The sources of contamination for the Hadnot Point water system were found to be hazardous waste and other materials, and an off-base dry cleaner was the likely source of contamination for the Tarawa Terrace water system. In 1989, the U.S. Environmental Protection Agency (EPA) placed both Camp Lejeune and the off-base dry cleaner (the source of PCE) on the National Priorities List (also known as Superfund sites). Since that time, several long-term actions to clean up the sources of contamination and to monitor and protect the base's drinking water have been implemented including the removal of contaminated soils and gasoline storage tanks, and the treatment of contaminated groundwater and soils (GAO, 2007).

AGENCY FOR TOXIC SUBSTANCES AND DISEASE REGISTRY RESPONSE TO CONTAMINATION

Concerns about possible adverse health effects associated with exposure to such solvents as TCE and PCE, and to various solvent mixtures led to a variety of activities, including health studies, claims against the federal government, and federal inquiries. From 1991 to 1997, the Agency for Toxic Substances and Disease Registry (ATSDR), which is part of the Centers for Disease Control and Prevention, conducted a public health assessment at Camp Lejeune that included a reconstruction of the estimated contaminant levels in the drinking water systems. In 2006, ATSDR estimated that well contamination from the off-base dry cleaner began as early as 1957. More recent ATSDR assessments indicate that it was most likely around August 1953 that the TCE contamination at Hadnot Point first exceeded the maximum contaminant levels (MCLs)[1] of 5µg/L for TCE, PCE, and benzene but that exceedances may have occurred as early as late 1948. PCE concentrations were found to have exceeded the MCL of 5µg/L for most of 1975–1985 with similar findings for the other contaminants (ATSDR, 2013) (see Table 1-1).

The exposure information compiled by ATSDR since the 1980s has been used in several epidemiologic studies of health effects on Camp Lejeune residents who were potentially exposed to contaminated water. These studies have examined a number of endpoints, including birth defects, adverse birth outcomes, cancer, and mortality (ATSDR, 2014).

In the birth defects investigation of 12,493 children born during 1968–1985 to mothers with residential exposure to contaminated drinking water at Camp Lejeune during pregnancy, ATSDR assessed exposure during the first trimester and looked for correlations between those exposures and childhood hematopoietic cancers, neural tube defects, and oral clefts. The telephone survey of parents, conducted between September 1999 and January 2002 (76% response rate), indicated that the number of children with birth defects was small (Ruckart et al., 2013). This study is discussed in more detail in Chapter 3.

For its first mortality study, ATSDR compared deaths among 154,932 marine and Navy personnel who began active duty between 1975 and 1985 and who served at Camp Lejeune with a similar group of 154,969 who served at Camp Pendleton, California, but not at Camp Lejeune, during this time. As of 2008, about 6% (8,964 at Camp Lejeune and 9,365 at Camp Pendleton) of both cohorts had died. When compared with general U.S. mortality rates, most of the standardized mortality ratios (SMRs) for the military cohort were below 1.0 indicating an expected healthy veteran effect (Bove et al., 2014a). A second mortality study compared 4,647 civilian employees who worked on base during that time period with 4,690 civilian employees at Camp Pendleton. As of 2008, about 14% (654) of the civilian cohort at Camp Lejeune had died (Bove et al., 2014b). The authors concluded that long-term follow up is necessary for a more complete assessment. Detailed results of both studies are discussed, where appropriate, in the following chapters.

Although the two studies used similar methodologies, there were several differences between the civilian and military Camp Lejeune cohorts. The proportion of women was much greater among the civilians (57.2% vs 5.2%); the median age of the civilians was about 10 years greater than that of the military cohort (58 vs 49 years old at the end of the follow up in 2008); the median months employed or served was less in the civilians (29 vs 36 months); and the class of employment also differed, with 69.7% of the civilians performing white-collar jobs while only 3.6% of the military cohort were officers. While some outcomes have already shown potentially higher

[1] MCLs represent the highest level of contaminant allowed in drinking water as set by EPA. It is reasonable to expect small amounts of these contaminants in drinking water and they do not necessarily pose a health risk. EPA MCLs represent determinations of acceptable risk based on scientific literature and scientific opinion for both cancer and non-cancer effects. MCLs are set as close as feasible to the maximum contaminant level goal for that contaminant using the best available treatment technology (e.g., analytical detection limits) and taking cost into consideration. A maximum contaminant level goal is the level of a contaminant in drinking water below which there is no known or expected risk to health. MCLs are enforceable standards.

TABLE 1-1 Maximum Contaminant Levels and Maximum Estimated Levels of Contaminants Found in Camp Lejeune Drinking Water

Contaminant	Maximum estimated level in finished water (µg/L)		Maximum contaminant level (µg/L)
	Tarawa Terrace	Hadnot Point	
Trichloroethylene (TCE)	5–6	783	5
Tetrachloroethylene (PCE)	180	39	5
Trans 1,2-dichloroethylene (1,2-tDCE)	(<100)	435	100
Vinyl chloride	2	67	2
Benzene	N/A	12	5

NOTE: Maximum contaminant levels were all set by EPA in 1989 or later—several years after the exposure occurred at Camp Lejeune.
SOURCE: Hadnot Point and Holcomb Boulevard estimates from ATSDR (2013); maximum estimates for Tarawa Terrace from ATSDR (2007).

incidences in Camp Lejeune military personnel and civilians, it is important to note that members of these cohorts were not old enough at the time of follow up to have developed many of the outcomes of interest, let alone to have died from them. Nonetheless, these studies are informative because they provide the best information available on the populations of interest.

ATSDR is currently conducting two additional studies of Camp Lejeune residents (F. Bove, ATSDR, personal communication, July 24, 2014), has recently published a third study, and has proposed a fourth one. The first is a health survey (also referred to as a morbidity study) of Camp Lejeune military personnel, their dependents, and civilians that is designed to study how contaminated water may have affected subjects' health. The target population for the survey includes about 300,000 people who lived or worked at Camp Lejeune or Camp Pendleton (comparison group) before 1986. The 26-page survey began in 2011 and includes questions about more than 20 types of cancers and other diseases, as well as opportunities for open-ended responses (ATSDR, 2012). Self-reported diseases of interest are confirmed by medical records and cancer registries. Results are expected to be published in 2015.

A second ongoing study compares cases of male breast cancer to other cancer cases not known to be related to solvent exposure using the Veterans Affairs Central Cancer Registry. This case-control study of all marines seeks to determine if cases were more likely than controls to have been exposed to contaminated water while residing at Camp Lejeune. Publication of this study is anticipated in 2015.

A recently published study (Ruckert et al., 2014) looks into the effects of contaminated water at Camp Lejeune on pregnant women, with a specific focus on adverse birth outcomes (a re-analysis of data first published by ATSDR in 1998). The new analysis includes updated information about exposure based on modeling done as recently as 2013.

ATSDR is also proposing a study to assess cancer incidence among military personnel and civilians who resided at Camp Lejeune. ATSDR's proposal builds on the cohort identified in the mortality study described earlier (Bove et al., 2014a) by linking to state cancer registries and the Veterans Affairs Central Cancer Registry. As of July 2014, the methods and protocol were under development. Final results are not likely to be published for several years.

NATIONAL RESEARCH COUNCIL REPORT

In 2009 the National Research Council (NRC) released *Contaminated Water Supplies at Camp Lejeune: Assessing Potential Health Effects* in response to a request from Congress to independently assess potential health outcomes associated with past exposure to contaminated water at Camp Lejeune. The committee focused its attention on the toxicologic and epidemiologic literature regarding the effects of TCE, PCE, and solvent mixtures

BOX 1-1
Categories of Association

Sufficient evidence of a causal relationship: Evidence is sufficient to conclude that a causal relationship exists between exposure to combustion products and a health outcome in humans. The evidence fulfills the criteria for sufficient evidence of a causal association and satisfies several of the criteria used to assess causality: strength of association, dose–response relationship, consistency of association, temporal relationship, specificity of association, and biologic plausibility.

Sufficient evidence of an association: Evidence is sufficient to conclude that there is a positive association. That is, a positive association has been observed between exposure to combustion products and a health outcome in human studies in which bias and confounding could be ruled out with reasonable confidence.

Limited/suggestive evidence of an association: Evidence is suggestive of an association between exposure to combustion products and a health outcome in humans, but this is limited because chance, bias, and confounding could not be ruled out with confidence.

Inadequate/insufficient evidence to determine whether an association does or does not exist: The available studies are of insufficient quality, consistency, or statistical power to permit a conclusion regarding the presence or absence of an association between exposure to combustion products and a health outcome in humans.

Limited/suggestive evidence of no association: There are several adequate studies, covering the full range of levels of exposure that humans are known to encounter, that are mutually consistent in not showing a positive association between exposure to combustion products and a health outcome. A conclusion of no association is inevitably limited to the conditions, levels of exposure, and length of observation covered by the available studies. In addition, the possibility of a very small increase in risk at the levels of exposure studied can never be excluded.

SOURCE: IOM, 2000.

on Camp Lejeune residents, and similarly exposed populations. Health effects for which there was convergent toxicologic and epidemiologic information were of the most interest.

That committee assessed the associations between solvents and health outcomes that were found in the literature and also relied on information about such associations presented in an earlier Institute of Medicine (IOM) report, *Gulf War and Health, Volume 2: Insecticides and Solvents* (IOM, 2003). The IOM report reviewed the toxicologic and epidemiologic literature to assess the strength and nature of the association between exposure to solvents (primarily in occupational settings) and adverse health effects that might be seen in veterans who had served in the 1990–1991 Gulf War and had been exposed to solvents during deployment. Both the IOM and the NRC committees used five categories to represent the statistical association and strength of the evidence (see Box 1-1 for a description of each category of association).

Based on the evidence, the NRC committee identified the following health outcomes as having limited/suggestive evidence of an association with exposure to TCE, PCE, or solvent mixtures.

- Cancers of the breast, bladder, kidney, esophagus, and lung were associated with TCE or PCE exposure in the epidemiologic literature with the strongest support from toxicologic studies for kidney cancer associated with TCE exposure.

- Multiple myeloma, adult leukemia, and myelodysplastic syndromes were associated with chronic exposure to solvents in the epidemiologic literature.
- Hepatic damage and renal tubular-cell damage were seen in rodents exposed to high levels of TCE and PCE. Hepatic steatosis (fatty liver) and acute renal tubular necrosis were seen in epidemiologic studies of solvent exposures. Damage was associated with exposure to high levels of solvents, but not with chronic low-level exposures.
- Reproductive effects were less clear. Epidemiologic data suggested an association between female infertility and concurrent, but not previous, solvent exposure and between miscarriage and PCE exposure during pregnancy.
- Nervous system effects were seen in epidemiologic studies of inhaled solvents. Neurobehavioral effects were evident during exposure, but there was no evidence of such effects after exposure ceased. In toxicologic studies, nervous system effects were associated with high levels of TCE (e.g., central nervous system depression, attention deficits, alterations in visual evoked potentials), with high levels of PCE (e.g., anesthetic effects), and with low levels of PCE (e.g., changes in behavior and neurochemical markers).
- Immune effects, manifested as chronic glomerulonephritis and scleroderma, were associated with solvent exposure in epidemiologic studies (specifically, scleroderma and TCE exposure), while toxicologic studies of TCE and PCE showed a variety of immune effects (skin sensitization, asthma, immunosuppression, and autoimmune disease for TCE; and allergic sensitization and immunosuppression for PCE).

No health outcomes related to exposure at Camp Lejeune were found to have sufficient evidence to support an association or causal relationship, nor did the NRC committee identify any outcomes with sufficient evidence of no association. Most of the health outcomes were categorized as having inadequate or insufficient evidence of an association. The committee noted that there were other outcomes for which there was insufficient evidence for the committee to make inferences about associations between the outcome and TCE, PCE, or other solvents, and it further noted that other health effects could not be ruled out simply because they had not been included on the list.

The NRC committee also found that while evidence suggested that the levels of exposure to TCE, PCE, and other solvents at Camp Lejeune were unlikely to have caused such health effects, the possibility that health effects were caused by water contamination could not be ruled out. The committee concluded that because of methodologic limitations additional research is unlikely to yield definitive results concerning whether and how residents were adversely affected by the water contamination.

The NRC's resulting list of 14 health outcomes (see Box 1-2) associated with TCE, PCE, or solvent mixtures with limited/suggestive evidence of an association was used by Congress and the U.S. Department of Veterans Affairs (VA) to inform further policy decisions.

BOX 1-2
Health Conditions Associated with Camp Lejeune Drinking Water in NRC (2009)*

Esophageal cancer	Myelodysplastic syndromes
Lung cancer	Renal toxicity
Breast cancer	Hepatic steatosis
Bladder cancer	Female infertility with concurrent exposure
Kidney cancer	Miscarriage with exposure during pregnancy
Adult leukemia	Scleroderma
Multiple myeloma	Neurobehavioral effects

* Associations were judged by the NRC committee to have "limited/suggestive evidence of an association."
SOURCE: NRC, 2009.

HONORING AMERICA'S VETERANS AND CARING FOR CAMP LEJEUNE VETERANS ACT OF 2012

In response to the needs of veterans and families exposed to contaminated water at Camp Lejeune, Congress passed the Honoring America's Veterans and Caring for Camp Lejeune Families Act of 2012, also known as the Janey Ensminger Act (P.L. 112-154). Section 102 provides health benefits to veterans and family members and designates VA as the last payer for services related to 14 eligible conditions listed in Box 1-2,[2] and it also adds non-Hodgkin's lymphoma[3] to the list of eligible conditions.

To be eligible for health benefits through VA's Camp Lejeune Program for Exposure to Chemically Contaminated Water, a veteran must have served on active duty at Camp Lejeune for 30 days or longer between January 1, 1957, and December 31, 1987. Eligibility criteria for the Camp Lejeune Family Member Program are similar, stipulating that family members must have resided at Camp Lejeune for 30 days or longer during the same time frame. (Note: The committee refers to both of these programs as the Camp Lejeune program in this report and does not distinguish between them.) Eligibility is "notwithstanding that there is insufficient medical evidence to conclude that such illnesses or conditions are attributable to such service" for veterans or to residence at Camp Lejeune for family members. Inclusion is extended to children of pregnant women who resided at Camp Lejeune (P.L. 112-154). However, reservists who trained at Camp Lejeune on active duty are not eligible (Walters, 2014).

Health benefits for the 15 conditions include hospital care, medical services, and reimbursement of co-payments for VA services. For family members receiving care outside of the VA system, VA will reimburse for hospital care or medical services as the last payer, that is, after all claims and coverage for payment, including other health insurance plans, have been resolved (Walters, 2014; P.L. 112-154). The act also stipulates that "hospital care and medical services may not be furnished . . . for an illness or condition of a family member that is found, in accordance with guidelines issued by the Under Secretary for Health, to have resulted from a cause other than the residence [at Camp Lejeune]."

DEPARTMENT OF VETERANS AFFAIRS CAMP LEJEUNE PROGRAM

In August 2012, VA took a series of steps to implement the Janey Ensminger Act. Since inception, eligible veterans have been enrolled in VA services as Priority 6, that is, veterans who are 0% service connected for a health problem.[4] VA experts from the Veterans Health Administration (VHA), the Veterans Benefits Administration, the VA Office of General Council, and the VA Office of Congressional and Legislative Affairs formed a task force to help implement the act (Walters, 2014).

Because VA does not regularly provide medical benefits to family members or reimbursement for their care outside VA, specific efforts were made to reach and communicate with these beneficiaries about the Camp Lejeune program. Infrastructure and educational changes included new information technology systems to track requests from eligible veterans and family members, and education for VHA medical staff and social workers about the program. Collaboration with the U.S. Department of Defense was also necessary to develop a system to determine administrative eligibility and to verify active-duty status or residence at Camp Lejeune for 30 days or longer between 1957 and 1987 (Walters, 2014). VA has drafted internal regulations and guidance for its clinicians regarding the Camp Lejeune program (Walters, 2014).

---

[2] This list of 15 is based on the NRC's 2009 report with a few changes. Adult leukemia became leukemia, exposure qualifications were removed for female infertility and miscarriage, and non-Hodgkin's lymphoma was added.

[3] Non-Hodgkin's lymphoma was found in the NRC report to have inadequate/insufficient evidence concerning whether an association exists; however, it was later identified by both EPA and the International Agency for Research on Cancer (IARC) as a health effect associated with TCE exposure (EPA, 2011; IARC, 2014).

[4] More information on criteria for VA priority groups may be found at http://www.va.gov/healthbenefits/resources/priority_groups.asp (accessed July 9, 2014).

Clinical Guidance

The *Guidance for VHA Staff: Honoring America's Veterans and Caring for Camp Lejeune Families Act of 2012, Section 102, Covered Clinical Conditions* (hereafter called "the guidance"; see Appendix B for the full text of the guidance) covers both clinical and procedural decisions that lead to determinations about an individual's eligibility for health benefits and about the coverage of specific medical services under the program. This guidance is not equivalent to VA's Clinical Practice Guidelines, which detail diagnostic procedures and treatment options for a variety of health outcomes. The guidance is restricted to the 15 conditions that are listed in the act.

The guidance takes the clinician though three decision points to determine an individual's eligibility for the program and benefits for a particular service. These are framed as three questions:

1. Does the applicant have a medical illness or condition specified in the law?
2. Is there another cause for the medical illness or condition?
3. Which treatments/bills are associated with the medical illness or condition?

Responding to all three questions requires clinical information and judgment. To assist clinicians, VA also developed a set of algorithms for several of the conditions to guide decisions of eligibility based on clinical information. The most recent draft guidance and algorithms (as of June 1, 2014) are provided in Appendix B.

The guidance is the result of instructions laid out by the legislation, VA policy decisions, and medical and scientific input. For example, while the guidance closely follows the conclusions reached in the NRC's report, VA made a policy decision to accept cancer and scleroderma diagnoses in eligible veterans regardless of other potential causes. VA also made the decision to provide health benefits for most medical costs (not just those directly related to a cancer diagnosis) during treatment for cancer because many cancer treatments affect the entire body.

COMMITTEE'S CHARGE

To ensure that the clinical guidance for the 15 covered medical conditions listed in P.L. 112-154 is "scientifically sound," VA asked the IOM to convene an ad hoc committee to review *Guidance for VHA Staff* and to make recommendations for its improvement. In addition, the committee was asked to address the following questions:

1. Based on the latest scientific literature and the committee's review, describe the medical conditions that result from "renal toxicity" due to solvent exposures.
2. Based on the latest scientific literature and the committee's review, characterize the "neurobehavioral effects" as mandated for coverage in the law.

APPROACH

To address its task, the IOM convened a committee of 12 experts with experience in clinical medicine, occupational and environmental health, epidemiology, toxicology, neurology, and nephrology.

The committee held two open meetings. At the first, VA presented the charge to the committee and discussed its preparation of guidance. At the second session, further discussions were held with VA to clarify the scope and use of the guidance document.

The committee followed the path of the 2009 NRC committee and concentrated on the adverse effects associated with the primary solvents found in the drinking water at Camp Lejeune—TCE and PCE. It also considered studies of mixed solvents where appropriate. The committee noted that, in general, studies of mixed solvents did not allow for adverse effects to be attributed to an individual chemical in a mixture of many and that those mixtures often contained well-characterized toxic substances such as toluene, which was not a contaminant of concern at Camp Lejeune. Although contaminants other than TCE and PCE—such as benzene, toluene, and vinyl chloride—were present in the drinking water at Camp Lejeune, they were generally found at very low concentrations and not in all samples.

For the renal and neurobehavioral endpoints, the committee conducted literature searches in TOXLINE for epidemiologic and toxicologic studies on TCE, PCE, benzene, vinyl chloride, and mixed solvents published in 2008 or later (U.S. National Library of Medicine, 2014). Searches for epidemiologic studies on the various endpoints were also conducted in PubMed and in Google Scholar. Additional targeted searches were conducted to address committee needs and identified gaps. Searches were conducted for authoritative reviews, including EPA Toxicological Reviews, ATSDR Toxicological Profiles, and IARC Monographs published in 2008 or later. The searches identified literature published though the summer of 2014.

The committee reviewed the available literature to identify possible renal and neurobehavioral endpoints; no endpoints were ruled out a priori. The committee adopted a rule that in order to be considered a possible effect, a renal or neurobehavioral effect must be reported with statistical significance[5] in at least one relatively well-designed study, or otherwise have enough weight of evidence to be considered a possible effect. For the other outcomes listed in the legislation, the committee reviewed recent literature and previous assessments as well as additional information about each clinical entity to determine whether or not the guidance and algorithms were scientifically sound and to see what, if any, changes might improve them.

The committee was aware of several issues while reviewing the literature and making its findings and recommendations. In cases where the weight of the evidence was sparse but showed a positive association or was equivocal and expert judgment was used in making the finding, the committee gave the benefit of the doubt to the veteran and family members.

Because the only exposure of concern in this report is whether the veteran or family member resided at Camp Lejeune for at least 30 days during the period of contamination covered by the act, the committee considered all the epidemiologic and toxicologic literature on renal and neurobehavioral effects regardless of issues such as dose–response; timing (e.g., in utero, early childhood), route, and duration of exposure; and whether the primary exposure was to TCE, PCE, or another Camp Lejeune drinking water contaminant. As noted in the NRC report (2009), given the numerous limitations in the data and methodologic shortcomings, "only crude estimates of contaminants in the water supply can be obtained," and only qualitative estimates of dose were applicable. The committee also recognized that many factors can affect the etiology and presentation of a health condition. This is of particular concern as many of the health conditions in the VA clinical guidance require that the clinician determine whether the health condition was caused by something other than the patient's residence at Camp Lejeune during the time of contamination. VA has made a policy decision that clinicians do not have to consider exclusionary factors for any of the cancers or for scleroderma. Because these are policy decisions and not necessarily based on scientific evidence, the committee did not comment on the validity of these decisions.

Furthermore, the committee recognized that in addition to the possible exclusionary conditions presented in the guidance and algorithms, there are many other risk factors that may influence the development and presentation of a health condition, including life style, occupational exposures, genetics, other health conditions, and the duration and extent of a person's exposure to the Camp Lejeune contaminants. Although such factors can directly and indirectly affect the etiology of a health condition, it was not possible for the committee to comment on all possible risk factors that may affect the population of concern, nor are they discussed in the guidance. The committee also recognized that given the number of factors that may contribute to the etiology of a condition, it may be impractical to determine the likelihood of a condition resulting from exposure to Camp Lejeune drinking water alone and exclude other well-defined contributory factors such as diabetes and chronic kidney disease, particularly in a clinical setting. The committee notes that for the most part, the effects of synergistic, additive, inhibitory, and other interactions between the contaminants found at Camp Lejeune and other risk factors and the health outcomes listed in the legislation are unknown.

[5] Statistical significance may be represented by a confidence interval or a p-value. If the 95% confidence interval for a risk estimate (such as a risk ratio [RR] or odds ratio [OR]) includes 1.0, the association is not considered to be statistically significant; however, if the interval does not include 1.0, the association is said to be statistically significant with an alpha error (likelihood that the association is due to chance) of 5% (that is, $p < 0.05$). The confidence interval is considered to be the range in which there is a 0.95 probability (95% chance) that the true value falls.

ORGANIZATION OF THE REPORT

This report addresses the committee's three primary tasks. Chapter 2 provides a review of the literature and the VA guidance with regard to the characterization of the renal toxicity endpoints and the clinical outcomes that may be associated with exposure to the water contaminants at Camp Lejeune. Chapter 3 provides a similar analysis for neurobehavioral effects that may result from exposure to TCE, PCE, and other solvents found at Camp Lejeune. Clinical considerations as presented in the VA guidance for the other 13 conditions listed in the Camp Lejeune legislation are reviewed in Chapter 4. Although these three chapters assess the scientific soundness of the VA clinical guidance, the committee also sought to provide VA with recommendations for improving the accuracy and usability of the guidance. Chapter 5 presents the committee's discussion of VA's approach to decision making in the guidance and future considerations for using and updating the guidance. Finally, the report contains three appendices: Appendix A provides short biographical sketches of the committee members; Appendix B is the draft *Guidance for VHA Staff: Honoring America's Veterans and Caring for Camp Lejeune Families Act of 2012, Section 102, Covered Clinical Conditions* and the accompanying algorithms that the committee reviewed for this report; and Appendix C contains the relevant sections of the Honoring America's Veterans and Caring for Camp Lejeune Families Act of 2012.

REFERENCES

ATSDR (Agency for Toxic Substances and Disease Registry). 1998. *Volatile organic compounds in drinking water and adverse pregnancy outcomes United States Marine Corps Base, Camp Lejeune, North Carolina.* Available at www.atsdr.cdc.gov/HS/lejeune (accessed October 23, 2014).

ATSDR. 2007. *Analyses of groundwater flow, contaminant fate and transport, and distribution of drinking water at Tarawa Terrace and vicinity, U.S. Marine Corps Base Camp Lejeune, North Carolina: Historical reconstruction and present-day conditions. Chapter A: Summary of findings.* Atlanta, GA: Centers for Disease Control and Prevention.

ATSDR. 2012. *ATSDR health survey of Marine Corps personnel and civilians.* Available at www.atsdr.cdc.gov/sites/lejeune/health_survey.html (accessed October 23, 2014).

ATSDR. 2013. *Analyses and historical reconstruction of groundwater flow, contaminant fate and transport, and distribution of drinking water within the service areas of the Hadnot Point and Holcomb Boulevard water treatment plants and vicinities, U.S. Marine Corps Base Camp Lejeune, North Carolina. Chapter A: Summary and findings.* Atlanta, GA: Centers for Disease Control and Prevention.

ATSDR. 2014. *Health study activities frequently asked questions (FAQs).* Available at www.atsdr.cdc.gov/sites/lejeune/qa_healthstudyactivities.html#5 (accessed October 23, 2014).

Bove, F. J., P. Z. Ruckart, M. Maslia, and T. C. Larson. 2014a. Evaluation of mortality among marines and Navy personnel exposed to contaminated drinking water at USMC Base Camp Lejeune: A retrospective cohort study. *Environmental Health* 13:10.

Bove, F. J., P. Z. Ruckart, M. Maslia, and T. C. Larson. 2014b. Mortality study of civilian employees exposed to contaminated drinking water at USMC Base Camp Lejeune: A retrospective cohort study. *Environmental Health* 13:68.

EPA (U.S. Environmental Protection Agency). 2011. *Toxicological review of tricholorethylene (CAS no. 79-01-6) in support of summary information on the Integrated Risk Information System (IRIS).* EPA/635/R-08/011F. U.S. Environmental Protection Agency.

GAO (Government Accountability Office). 2007. *Defense of health care: Activities related to past drinking water contamination at Marine Corps Base Camp Lejeune. GAO-07-276.*

IARC (International Agency for Research on Cancer). 2014. Trichloroethylene. In *IARC Monographs on the Evaluation of Carcinogenic Risks to Humans, Vol. 106.* Lyon, France: International Agency for Research on Cancer.

IOM (Institute of Medicine). 2000. *Gulf War and health, Volume 1: Depleted uranium, pyrodostigmine bromide, sarin, vaccines.* Washington, DC: National Academy Press.

IOM. 2003. *Gulf War and health, Volume 2: Insecticides and solvents.* Washington, DC: The National Academies Press.

NRC (National Research Council). 2009. *Contaminated water supplies at Camp Lejeune: Assessing potential health effects.* Washington, DC: The National Academies Press.

Ruckart, P. Z., F. J. Bove, and M. Maslia. 2014. Evaluation of contaminated drinking water and preterm birth, small for gestation, age, and birth weight at Marine Corps Base Camp Lejeune, North Carolina: A cross-sectional study. *Environmental Health* 13:99.

U.S. Marine Corps. Undated. *Marine Corps Base Camp Lejeune: About*. Available at www.lejeune.marines.mil/About.aspx (accessed August 15, 2014).

U.S. National Library of Medicine. 2014. *TOXLINE*. Available at toxnet.nlm.nih.gov/cgi-bin/sis/htmlgen?TOXLINE (accessed October 23, 2012).

Walters, T. 2014. Charge to Committee: Institute of Medicine Committee to Define Clinical Terms in P.L. 112-154 & Review Clinical Guidelines. Washington, DC, May 15, 2014.

2

Characterization of Renal Toxicity

In 2009, the National Research Council (NRC) Committee on Contaminated Drinking Water at Camp Lejeune reviewed the scientific evidence on the association between renal toxicity and exposure to the solvents found in the drinking water at Camp Lejeune. That committee began its work by reviewing a 2003 Institute of Medicine (IOM) report on solvent exposure and possible health effects, and it also reviewed new toxicologic and epidemiologic studies published from 2003 through 2008. The NRC report included an in-depth examination of both human and animal studies of the renal toxicity induced by exposure to two solvents: trichloroethylene (TCE) and perchloroethylene (1,1,2,2-tetrachloroethylene, or PCE). The few animal data available on the kidney toxicity of 1,2-dichloroethylene, 1,1-dichloroethylene, methylene chloride, benzene, and vinyl chloride were also considered. In general, these studies found that high-dose solvent exposures were necessary to produce acute renal effects and that the effects of such exposures were variable among species. The 2009 NRC report concluded that there was limited/suggestive evidence of an association between exposure to mixed solvents and renal toxicity.

Few new data have been published concerning the renal toxicity of the Camp Lejeune drinking water contaminants, other than TCE and PCE, since 2008. This chapter reviews new information, which has come primarily from animal and human toxicity studies with TCE and PCE, and it discusses that new information in the context of the previous conclusions concerning the renal toxicity of the contaminants.

The chapter begins with a summary of the previous assessments of the association between solvent exposure and renal toxicity. A description of possible mechanisms for this toxicity follows. The committee then reviews recent animal and epidemiologic studies of the associations between solvents and renal effects and it draws conclusions about the association between solvent exposure and specific renal effects. Finally, the chapter discusses the U.S. Department of Veterans Affairs (VA) clinical guidance and algorithm K for renal toxicity for the Camp Lejeune program and suggests a modified algorithm. Kidney cancer, one of the cancers covered by the Janey Ensminger Act (P.L. 112-154), is discussed in Chapter 4 in the section on "Cancers and Related Conditions."

PREVIOUS ASSESSMENTS

This committee conducted an extensive literature search for new information published between 2008 and 2014 to help define the renal toxicity that might result from exposure to the solvents found in the drinking water at Camp Lejeune. The studies reviewed in the previous IOM and NRC reports were not reassessed but rather were used to provide the appropriate background for interpreting new evidence on the renal toxicity of the Camp

Lejeune drinking water contaminants to humans. The 2003 IOM and 2009 NRC committees' evaluations of and conclusions about human renal toxicity based on the available epidemiological studies are summarized below.

2003 IOM Report

In 2003, the IOM released a report that assessed the long-term health consequences that might occur in veterans of the 1990–1991 Gulf War who may have been exposed to solvents during their deployment to the Persian Gulf. Many of the studies reviewed were of occupational exposures to a variety of solvents and solvent mixtures. Studies of the effects of short-term and long-term solvent exposure on renal function below the threshold of clinical disease provided some support for an association between exposure to high concentrations of solvents and acute tubular necrosis. A series of case-control studies that evaluated chronic glomerulonephritis, an immune-mediated disease, in relation to nonspecific solvent exposure provided inconsistent evidence of an association; however, several reasonably strong studies showed dose–response gradients. One large study (Steenland and Palu, 1999) reported a reasonably strong association between an exposure to solvents used for cleaning and degreasing and end-stage renal disease (ESRD), and the study by Porro et al. (1992) reported an association between "degreasing agents" and ESRD. The IOM study concluded that there was limited/suggestive evidence of an association between exposure to solvent mixtures and chronic glomerulonephritis (see Chapter 1, Box 1-1 for a description of the categories of association). It noted that although several studies had addressed the effect of solvent exposure on indicators of renal function, these studies used various magnitudes of exposure and the quality of the exposure assessments varied. None of the studies addressed TCE or PCE directly.

2009 NRC Report

The 2009 NRC report did not identify any new studies of solvent exposure and glomerulonephritis. A large, occupational cohort study of aircraft-maintenance employees did find a nearly two-fold increase in the odds of ESRD (OR [odds ratio] = 1.91, 95% CI [confidence interval] 1.08–3.38) among workers exposed to TCE but not among those exposed to PCE (Radican et al., 2006). A study of renal function in electronics workers who were exposed to TCE (mean concentration 32 ppm [parts per million]; range 0.5–252 ppm) showed decrements in renal function in the clinically normal range. While a small effect on renal function was observed, these effects were not related to extent of exposure and led the authors to conclude that there was no evidence of kidney toxicity under the exposure conditions studied (Green et al., 2004).

The report concluded "that there continues to be limited/suggestive evidence of an association between mixed solvent exposure and chronic glomerulonephritis but inadequate/insufficient evidence to determine whether an association exists specifically between TCE or PCE and chronic glomerulonephritis" (NRC, 2009, p. 158). The report further concluded that animal studies had found high concentrations of TCE and PCE to result in renal tubular cell damage, and that epidemiologic studies provided limited/suggestive evidence of an association between chronic high-level exposures to solvents (but not chronic low-level exposure) and acute renal tubular necrosis (NRC, 2009).

Mechanism of Nephrotoxicity

TCE has been assessed in lifetime carcinogenic bioassays in rats and mice by the National Cancer Institute (NCI, 1976) and the National Toxicology Program (NTP, 1988, 1990); the NTP has also assessed the lifetime carcinogenicity of PCE (NTP, 1986). Those studies found kidney cancer in both species, which stimulated research to characterize the toxicity and potential mechanisms, or modes of action, for both chemicals. Several epidemiological studies were undertaken to characterize the effects on humans of occupational exposures to these chemicals.

The effects of TCE and PCE on kidney function have both been studied in animal models, primarily rodents. Such studies allow measurements of multiple aspects of renal function with varying exposure durations and help elucidate the production of TCE and PCE metabolites. In vitro studies are particularly useful for assessing potential toxic pathways, including the generation of reactive metabolites, the disruption of cellular energy processes, and the production of oxidative stress. These studies have shown that acute exposure to high doses of TCE causes tubular necrosis localized to the proximal sections of the nephron, which results in impaired reabsorption of solutes, including glucose, protein, and water (Chakrabarti and Tuchweber, 1988). Intrarenal control mechanisms constrict blood flow to the glomeruli of the damaged tubules, decreasing the glomerular filtration rate (GFR). Chronic (2-year) and subchronic (13-week) exposures to high doses of either TCE or PCE (generally 300–1000 mg/kg/d) cause kidney pathology, reported as cytomegaly, karyomegaly, and necrosis of the tubular epithelium, particularly in the inner cortical tubular area. In these studies, pathology was determined immediately after exposure.

Studies using experimental animals and cultured cells (including human) have shown that, as with other solvents, neither TCE nor PCE itself is toxic, but rather each is metabolized to chemically reactive intermediates. TCE and PCE are different from many toxicants in that both oxidative and conjugation reactions are required to produce toxic metabolites. A simplified pathway for TCE metabolism is shown in Figure 2-1.

TCE metabolism was reviewed in Lash et al. (2000) and NRC (2006). TCE is metabolized by cytochrome P450 to oxidative metabolites and, by conjugation, to glutathione (see Figure 2-1). The oxidative metabolites are chloral, chloral hydrate, trichloroethanol (TCOH), and trichloroacetic acid (TCA). The glutathione conjugate, dichlorovinyl glutathione (DCVG), is metabolized by γ-glutamyltranspeptidase (γGTP) to dichlorovinylcysteine (DCVC). DCVC has multiple fates. It can be further metabolized to dichlorovinylthiol (DCVT) by cysteine conjugate β-lyase; to dichlorovinylcysteine sulfoxide (DCVCS) by flavin-containing monooxygenase 3 (FMO3), or by N-acetyltransferase (NAT) to N-acetyl dichlorovinylcysteine (NAcDCVC) and then to N-acetyl dichlorovinylcysteine sulfoxide (NAcDCVCS) by cytochrome P4503A (CYP3A). The toxicity of all these metabolites to kidney

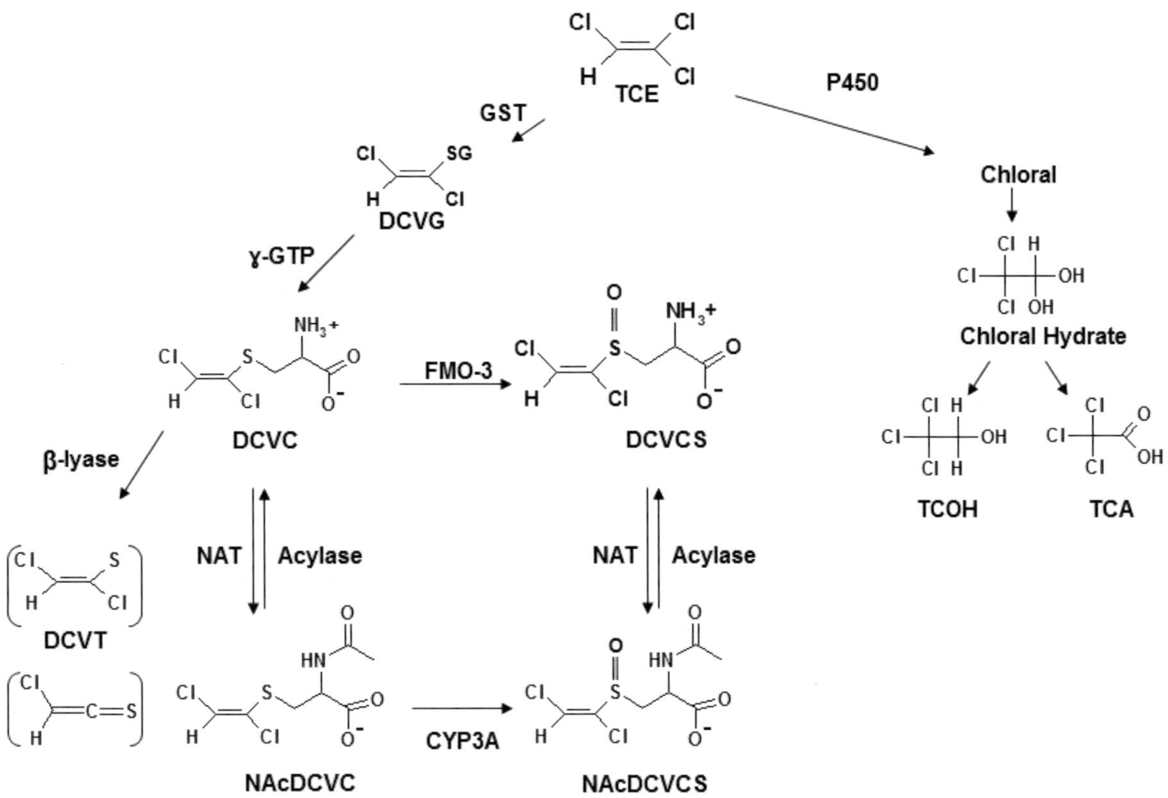

FIGURE 2-1 Metabolic pathway for TCE.
SOURCE: Adapted from NRC, 2006.

cells has been well established in both in vivo and in vitro studies; however, DCVC and its subsequent metabolites account for the majority of the renal toxicity.

PCE metabolism was also reviewed by NRC (2010). PCE is metabolized by the same routes as TCE to reactive and toxic metabolites, specifically trichlorovinylcysteine (TCVC), N-acetyl trichlorovinylcysteine (NAcTCVC), and then N-acetyl trichlorovinylcysteine sulfoxide (NAcTCVCS).

REVIEW OF RECENT ANIMAL STUDIES

Metabolism

Recent research indicates that different TCE metabolites are selective for different subregions within the rat kidney, and therefore different segments of the nephron. Irving and colleagues (2013) administered specific metabolites to rats. They reported that NAcDCVC increases urine output somewhat (0.32 dL, compared with 0.20 dL for saline control), whereas DCVCS causes a dramatic drop in urine volume (0.05 dL, 2 animals were anuric). Histopathologically, the N-acetylated derivatives damage the corticomedullary junction whereas DCVCS targets the outer cortex. Massive glycosuria was observed with the N-acetylated derivatives of DCVC and DCVCS (approximately150 mg/24 hr, on average) whereas glycosuria was only slightly increased (10 mg/24 hr) following DCVC. These results indicate that different metabolites of TCE have relative selectivity for different segments of the proximal tubule.

Individual variation in the abundance of the enzymes involved in TCE metabolism—or the presence of other agents such as drugs that alter metabolism by these enzymes—would be expected to result in different patterns of tubular toxicity. Human variation in these enzymes is known to occur. FMO3 is a polymorphic drug-metabolizing enzyme found in the liver and, to a lesser extent, in the kidneys (Yamazaki and Shimizu, 2013).

NAT exists as both cytosolic and microsomal forms and both metabolize nephrotoxic halogenated cysteine conjugates (Altuntas and Kharasch, 2002). Rats, mice, and humans each have two forms of cytosolic NAT, denoted as 1 and 2 (National Center for Biotechnology Information Entrez Gene database). The substrates metabolized by each isoform are different in each species, so that a given substrate may be metabolized by, for example, isoform 1 in rats and isoform 2 in humans. In humans, NAT1 and NAT2 both have rapid phenotypes, and each is associated with an increased risk of renal toxicity upon exposure to drugs or other chemicals (Walker et al., 2009). NAT8 is the microsomal form. Recent studies have associated a mutation in NAT8 (rs 13538) with decreased kidney function in a resequencing study focused on NAT8 (Juhanson et al., 2008) and in genome-wide association studies (Kottgen et al., 2010). Veiga-da-Cunha and colleagues (2010) used HEK293T cells to express NAT8 and rs 13538 mutant (phenylalanine replaced by serine at position 143). They reported decreased activity, due to decreased expression, of the mutated enzyme. Altuntas and Kharasach (2002) reported N-acetylation of cysteine conjugates by both cytosolic and microsomal fractions of human kidney to be highly variable, across twenty samples, and to differ with substrate. Thus, the signs and symptoms of renal toxicity will vary depending on an individual's metabolic pattern, which implies that a finding in rats may or may not reliably predict acetylation in humans.

Excretion

Using an in vitro model of renal tubular epithelium (cell culture) with cells expressing Mrp2, Tsirulnikov et al. (2010) showed that NAcDCVC undergoes basolateral to apical transport. Mrp2 is a member of the adenosine triphosphate (ATP)-binding cassette class of active transporters that perform the secretion of intracellular NAcDCVC across the apical membrane into the tubular urine, that is, the second step in the movement of TCE metabolites from blood to urine. Mrp2 transported most of the NAcDCVC (Tsirulnikov et al., 2010). Human MRP2 is known to have genetic variants (see, for example, OMIM, 2014), and it is subject to induction by substrates such as rifampin, dexamethasone, and phenobarbital and to inhibition by substrates such as vinca alkaloids, anthracyclines, and cisplatin. Thus, individuals may differ in their ability to transport intracellular NAcDCVC out of the cell. Individuals with less capacity to transport NAcDCVC out of the cell will have greater exposure to this toxic agent, resulting in increased toxicity to the tubular cells.

Guinea Pig Sensitization Model

Yu and co-workers (2012) assessed the renal effects of a TCE challenge in guinea pigs that had been previously sensitized to TCE (via injection of Freund's complete adjuvant and TCE). This experiment was motivated by the observation of an "occupational medicamentosa-like dermatitis" in Chinese workers that was sometimes fatal. The TCE challenge to sensitized guinea pigs (those that exhibited allergic reactions to intradermal injection of TCE) produced histopathological lesions and an impairment of renal function (increased blood urea nitrogen, increased excretion of proteins in urine, including beta-2-microglobulin, alkaline aminopeptidase, and gamma-glutamyl transpeptidase). Histopathological tubular changes included the necrosis of cells, the loss of the brush border, mitochondrial damage (vacuolar swelling), and the fusion of foot processes in the glomeruli. These results suggest that in the presence of Freund's complete adjuvant exposure to TCE can lead to an immune response upon subsequent challenge. Freund's complete adjuvant is an emulsion of antigen in mineral oil, and it is used because it is effective in stimulating cell-mediated immunity, thus enhancing the biological response so that events occur more frequently and can be better studied. The applicability of this research model to human exposure via water consumption, without the stimulating effect of Freund's adjuvant is not known. Thus, TCE's glomerular effects are unknown.

Conclusions from Animal Studies

The metabolic pathways of TCE and of PCE have been well characterized in animal models. Humans have similar enzymes and, in general, produce the same metabolites as in the animal models. Some of these enzymes, however, have polymorphisms in the human population. The secretion of metabolites into urine is mediated by MRP2, known to be affected by drugs in common use (see earlier section on "Excretion"). The variability in the processes involved in producing and eliminating the TCE and PCE metabolites would be expected to result in variability in the magnitude of responses after exposure to these chemicals. Furthermore, each reactive metabolite has a different toxicity so the observed effects would also be expected to vary between and among humans and animals. Although studies support the existence of this intra- and inter-species variation in the toxic response to exposure to TCE or PCE, the variation is not sufficiently well characterized to allow easy extrapolation from animals to humans and vice versa. Finally, in spite of the number of animal studies on the renal toxicity of TCE and PCE, neither the 2009 NRC committee nor this committee identified any animal studies with exposures similar to those that occurred at Camp Lejeune, that is, that assessed long-term renal effects following short-term exposure to the solvents as either immature or adult animals.

REVIEW OF RECENT EPIDEMIOLOGICAL STUDIES

In reviewing the recent literature related to the renal effects of the drinking water contaminants at Camp Lejeune, the committee identified three new epidemiological studies on TCE and one on PCE. In addition one review assessed the updated Integrated Risk Information System (IRIS) report on the human health risk assessment of TCE, prepared by the U.S. Environmental Protection Agency (EPA). These studies are reviewed below.

Calvert et al. (2011) examined the incidence of and mortality from ESRD in workers exposed to PCE only or PCE plus other solvents (most likely Stoddard's solvent). This study was the third mortality update (as of 1977) on a cohort of 1,704 dry-cleaning workers in four U.S. cities (Chicago, Detroit, New York, and San Francisco/Oakland). Although ESRD from all causes in the entire cohort was nonsignificantly elevated (standardized incidence ratio [SIR] 1.34, 95% CI 0.90–1.91), hypertensive ESRD morbidity was elevated in the entire cohort (SIR = 1.98, 95% CI 1.11–3.27) as well as those employed in PCE-only dry-cleaning establishments for greater than 5 years (SIR = 3.39, 95% CI 1.10–7.92). In addition, the underlying cause-of-death standardized mortality ratio (SMR) for PCE-only workers for acute glomerulonephritis, nephrotoxic syndrome, and acute renal failure was nonsignificantly increased (two deaths; SMR = 2.60, 95% CI 0.31–9.39), while mortality from chronic and unspecified nephritis, renal failure, and other renal sclerosis in the entire cohort was decreased nonsignificantly (two deaths, SMR = 0.42, 95% CI 0.05–1.52). The authors concluded that the increased risk for hypertensive ESRD among workers with PCE-only exposure, particularly for those workers with a longer duration of solvent exposure, supported the conclusion that PCE exposure, rather than lifestyle or socioeconomic factors, was associated with renal toxicity.

Earlier assessments of TCE had suggested that renal toxicity in humans might occur at high exposure levels on the basis of increased urinary protein excretion (NRC, 2009). Using a panel of novel sensitive nephrotoxicity markers, Vermeulen et al. (2012) examined renal toxicity in 80 Chinese factory workers exposed to TCE at concentrations below the permissible exposure level of 100 ppm (8-hour time-weighted average) set by the U.S. Occupational Safety and Health Administration (OSHA). The six factories selected for study used TCE in manufacturing processes but had no detectable levels of benzene, styrene, ethylene oxide, formaldehyde, or epichlorohydrin, and they had low to negligible levels of methylene chloride, chloroform, and PCE. The control set of factories from the clothing and food industries did not use TCE. For exposed workers, the average length of TCE exposure was 2 years, and the mean exposure concentration was 22 ppm; the 45 control workers had been employed an average of 2.3 years in their factories but had negligible TCE exposure. The authors found that one sensitive urinary protein biomarker of proximal tubular injury—kidney injury molecule-1 (KIM-1)—was significantly increased ($p = 0.01$) in exposed workers compared with controls. A second urinary biomarker—Pi-glutathione S transferase (Pi-GST)—was increased in exposed workers, but the increase did not reach statistical significance ($p = 0.09$). Pi-GST was considered to be a borderline indicator of renal toxicity. Other urinary markers of renal function (creatinine, alpha-GST, N-acetyl-β-D-glucosaminidase [NAG]) did not differ between TCE-exposed and unexposed workers. The authors concluded that renal toxicity, as evidenced by elevated KIM-1 and possibly Pi-GST urinary excretion, could occur at TCE concentrations below the OSHA exposure limit.

In 2011, EPA published an updated IRIS report on the human health risk assessment of TCE (EPA, 2011). The IRIS report contained a detailed review of literature concerning health risks associated with TCE exposure. It was noted that workers highly exposed to TCE exhibited evidence of tubular and possible glomerular damage, based on the presence of increased urinary excretion of α1-microglobulin, NAG, Pi-GST, or total protein. However, not all exposed groups exhibited the same patterns or degree of urinary protein excretion, and some workers were exposed to mixed solvents rather than TCE alone. These observations provide further support for the conclusions on proteinuria in the 2009 NRC report.

Bove et al. (2014a) reported on the causes of mortality for marine and Navy personnel who began service between 1975 and 1985 and were stationed at Camp Lejeune (n = 154,932) or at Camp Pendleton, California (n = 154,969) during those years; the mortality follow-up period was 1979–2008. The authors reported on mortality due both to kidney disease and to cancer. There were fewer deaths than expected from kidney diseases for both the Camp Lejeune cohort (SMR = 0.50, 95% CI 0.35–0.68) and the Camp Pendleton cohort (SMR = 0.52, 95% CI 0.37–0.71) compared with U.S. mortality rates, while there were more deaths than expected from kidney cancer in the Camp Lejeune cohort (SMR = 1.16, 95% CI 0.84–1.57), but fewer than expected in the Camp Pendleton cohort (SMR = 0.89, 95% CI 0.61–1.25). The risk of dying from kidney disease was the same for Camp Lejeune as for Camp Pendleton (hazard ratio [HR] = 1.00, 95% CI 0.63–1.63), and the Camp Lejeune cohort had a nonsignificant increased risk of dying from kidney cancer (HR = 1.35, 95% CI 0.84–2.16). The results suggest that the Camp Lejeune cohort did not have an increased risk of chronic renal toxicity leading to death or an increased risk of kidney cancer. The authors noted that 97% of the Camp Lejeune cohort was under the age of 55 and only 6% of the cohort had died of any cause by the end of the study; they cautioned that long-term follow up would be needed for a comprehensive assessment of the effects of exposure to the contaminated water at Camp Lejeune.

In a separate report, Bove et al. (2014b) compared the mortality of 4,647 civilian workers at Camp Lejeune during 1973–1985 with 4,690 nonexposed workers at Camp Pendleton during the same time; the mortality follow-up period was again 1979–2008. No significant kidney effects were found. There were fewer deaths than expected from kidney diseases in both the Camp Lejeune (SMR = 0.78, 95% CI 0.34–1.54) and Camp Pendleton (SMR = 0.50, 95% CI 0.22–1.00) cohorts compared with U.S. mortality rates. Deaths from kidney diseases were not associated with cumulative or average exposure to the drinking water contaminants. Deaths from kidney cancer were higher than normal (SMR = 1.30, 95% CI 0.52–2.67) at Camp Lejeune but not at Camp Pendleton (SMR = 0.82, 95% CI 0.30–1.80). The hazard ratios for deaths due to kidney diseases and kidney cancer at Camp Lejeune and Camp Pendleton both had nonsignificant effects (kidney diseases: 1.23, 95% CI 0.39–3.87; kidney cancer: 1.92, 95% CI 0.58–6.34). The authors noted that the study's limitations included the small numbers for most causes of death (e.g., each camp had seven deaths from kidney disease) and a potential exposure misclassification bias. Because only 14% of the Camp Lejeune and 18.5% of the Camp Pendleton subjects had died by the end of the study, the authors called for long-term follow-up studies to provide a comprehensive assessment of the effect of drinking water exposures at Camp Lejeune.

Summary of Human Studies

Calvert et al. (2011) found an increased risk of morbidity from solvent-induced hypertensive ESRD in workers with increasing years of exposure, but the number of workers with this outcome was small. The researchers failed to find any increase in mortality in the solvent-exposed workers from chronic and unspecified renal nephritis, renal failure, or other renal sclerosis. Studies by Bove et al. (2014a,b) did not demonstrate any increase in mortality from kidney disease in marine or Navy personnel or in civilian workers at Camp Lejeune, although longer-term follow up is needed. Nevertheless, these results do not support the existence of an increased risk of chronic renal toxicity leading to death in the military and civilian Camp Lejeune cohorts.

CONCLUSIONS FROM ANIMAL AND HUMAN STUDIES

Based on the evidence reviewed here and in previous reports (IOM, 2003; NRC, 2009) there appears to be strong evidence for an association between acute exposure to high levels of TCE or PCE and acute tubular toxicity in both rodents and humans, although humans metabolize these chemicals to a lesser extent and are thus more resistant to adverse effects. There is accumulating evidence that acute renal injury, as might occur soon after exposure, significantly increases the incidence of chronic kidney disease (CKD) many years later (Chawla et al., 2014). Such an effect could occur even if the acute injury were subclinical and thus not detected.

The evidence for an association of TCE or PCE with CKD is less clear, although there does appear to be an association between exposure to high levels of these solvents and ESRD (Calvert et al., 2011; Radican et al., 2006; Steenland et al., 1990). However, the documented levels of PCE and TCE in the drinking water at Camp Lejeune were much lower than those in the animal and human studies discussed here, and it is expected that the exposure duration (median of 36 months in the military cohort; see Bove et al., 2014a) would have been much shorter as well. There is no evidence for an increased incidence of CKD in those who served at Camp Lejeune during the time of the contaminated drinking water.

In humans, exposure to TCE and PCE occurs in complex settings where other etiologies of kidney disease may coexist. The present literature in humans does not permit one to distinguish whether TCE and PCE cause renal disease on their own, or interact with other causes of renal disease, enhancing their toxicity. Although the committee notes that kidney disease, including chronic glomerulonephritis and tubular necrosis, in those who resided at Camp Lejeune will likely be due to causes other than TCE or PCE exposure, it is not possible to rule out a role for solvent exposure. This is a common problem when seeking causes of kidney disease where there is no specific diagnostic histopathology. That kind of renal damage, should it occur, would present clinically as CKD, which describes any type of permanent kidney damage that may progress to ESRD (American Kidney Fund, 2012).

DISCUSSION OF GUIDANCE AND ALGORITHM

In this section, the committee assesses the VA clinical guidance and algorithm K on renal toxicity. The following discussion pertains to the proposed changes in the text of the guidance, the algorithm, and the annotations to the algorithm. Figure 2-2 shows a revised algorithm K that incorporates these suggested changes.

The VA's clinical guidance specifies that CKD, defined as a chronic decrease in kidney function, or proteinuria should be the clinical endpoints of concern for renal toxicity resulting from solvent exposure at Camp Lejeune. The committee finds CKD to be an appropriate endpoint to represent possible kidney damage potentially caused by exposure to contaminated water at Camp Lejeune.

The guidance asks first whether the patient has evidence of renal injury, when the onset of CKD occurred, and if the patient has other comorbid conditions. The clinician then assesses whether it is probable that the CKD

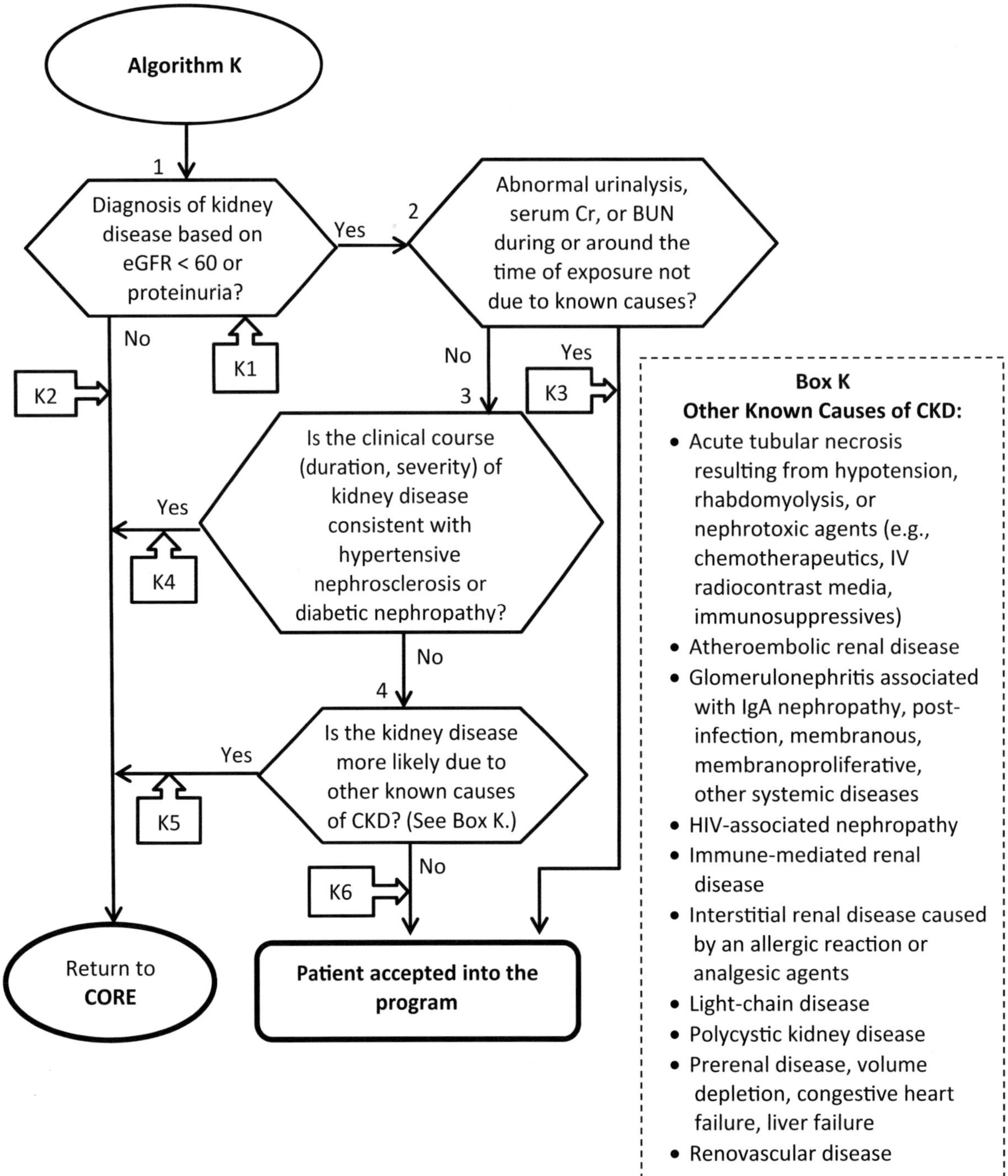

FIGURE 2-2 Revised algorithm K.

is attributable to a known cause other than solvent toxicity. If there is no evidence for another cause, CKD could be due to toxic exposure. The committee notes, however, that there are several reasons why there may be a lack of evidence of acute renal toxicity at the time of exposure: Renal toxicity did not occur; it did occur but the patient was asymptomatic and therefore there was no indication that the necessary laboratory tests should be conducted; or the tests were conducted but were not sensitive enough to detect mild disease. Thus, the committee finds that a patient should not be ineligible for the VA program because of a lack of documented evidence of kidney disease during or shortly after residence at Camp Lejeune.

If there is no history of acute renal injury around the time of residence at Camp Lejeune, the guidance asks clinicians to consider whether the patient has diabetes mellitus or hypertension (common causes of CKD) or other conditions associated with CKD (such as diabetic neuropathy, obstructive uropathy, hypertensive nephrosclerosis, sickle cell kidney disease, HIV-associated nephropathy, and drug-induced kidney disease) (see Table 2-1). If evidence for such conditions exists and the patient's course is consistent with those conditions (that is, the "renal disease is as likely as not associated" with those conditions), CKD should be attributed to those entities and the patient would not be accepted into the Camp Lejeune program. Conversely, if no other causes of CKD are probable, patients would be accepted into the program. Similarly, the guidance states that if the patient's disease is "atypical, in that the progression of their kidney disease is faster than expected, then exacerbation by TCE, PCE or other organic solvents in the contaminated water should be considered" and the patient should be admitted to the program.

Algorithm K addresses renal disease and reflects slightly different and more detailed information than the text in the guidance (see Table 2-1). The first step in the original algorithm directs a clinician to identify data for kidney damage such as eGFR (estimated glomerular filtration rate), serum creatinine, or other indicators of kidney

ANNOTATIONS FOR REVISED ALGORITHM K:

K1—Diagnosis of kidney disease: (1) Applicant has a history of renal toxicity or kidney disease concurrent with Camp Lejeune residence or shortly after the time of possible exposure to contaminated water at Camp Lejeune, or (2) applicant has evidence of chronic kidney disease (CKD).

The two most common causes of CKD are diabetes and hypertension. In most instances, it will be possible to identify the most likely cause of CKD using history, physical examination, laboratory testing, and imaging tests. A kidney biopsy should be considered for patients with nephrotic range proteinuria (urine to creatinine ratio > 3.5), particularly in the absence of diabetes, to determine the histopathology of the kidney disease. The decision to perform a kidney biopsy should be based on the need to provide optimal care to the patient.

K2—Applicant is still administratively eligible for the Camp Lejeune program but does not have evidence of renal toxicity as a covered condition.

K3—Applicant has a history of renal toxicity or kidney disease concurrent with Camp Lejeune residence or shortly after the time of possible exposure to contaminated water at Camp Lejeune. If this cannot be attributed to other known causes of kidney disease, it should be presumed that any subsequent kidney disease may be related to toxin exposure at Camp Lejeune, and the patient should be accepted into the program.

K4—Applicant has no history of renal toxicity or kidney disease concurrent with Camp Lejeune residence or around the time of possible exposure to contaminated water at Camp Lejeune. Applicant has evidence of kidney disease due to long-standing diabetes or refractory hypertension, which are common causes of kidney failure and are not related to exposure to the contaminants in the water at Camp Lejeune. Applicant does not have a covered condition and is not eligible for coverage by the Camp Lejeune program at this time.

K5—Applicant has no history of renal toxicity or kidney disease concurrent with Camp Lejeune residence or around the time of possible exposure to contaminated water at Camp Lejeune. Applicant has evidence of kidney disease consistent with a secondary condition that is not related to exposure to the contaminants in the water at Camp Lejeune. Current kidney disease is due to another cause other than exposure to contaminated water at Camp Lejeune. Applicant does not have a covered condition and is not eligible for coverage by the Camp Lejeune program at this time.

K6 [New]—Applicant has CKD of uncertain etiology, possibly related to exposure to contaminated water at Camp Lejeune. Applicant has a covered condition, renal toxicity, and is accepted into the Camp Lejeune program.

TABLE 2-1 Possible Exclusionary Causes of CKD as Given in the Clinical Guidance, the Original Algorithm K, and the Revised Algorithm K

Guidance Text	Algorithm (original)	Algorithm (revised)
Diabetes	Diabetes	Diabetic kidney disease
Hypertension/hypertensive nephrosclerosis	Hypertension	Hypertensive kidney disease
Obstructive uropathy	Urinary tract obstruction	Urinary tract obstruction
Sickle cell kidney disease		Sickle cell kidney disease
HIV-associated nephropathy		HIV-associated nephropathy
Drug-induced kidney disease	Acute tubular necrosis occurring in the setting of hypotension or nephrotoxic agents, such as radiocontrast, antibiotics, or chemotherapy drugs	Acute tubular necrosis occurring in the setting of hypotension, rhabdomyolysis, or nephrotoxic agents (e.g., chemotherapeutics, IV radiocontrast media, immunosuppressives)
	Acute interstitial nephritis, often due to drugs such as NSAIDs or antibiotics	Interstitial renal disease Allergic, analgesic agents
	Volume depletion, severe heart failure	Prerenal disease: volume depletion, congestive heart failure, liver failure
		Atheroembolic disease
		Glomerulonephritis
		IgA nephropathy, post-infection, membranous, membranoproliferative, associated with systemic diseases
		Immune-mediated renal disease
		Polycystic kidney disease
		Vasculitis
		Light chain disease

NOTE: HIV = human immunodeficiency virus; IgA = immunoglobulin A; IV = intravenous; NSAIDs = nonsteroidal anti-inflammatory drugs.

failure in the patient's medical record; these indicators are not specified in the original guidance. The committee finds that this step is unnecessary and therefore that it could be deleted from the algorithm.

The second step in the original algorithm (Box 2 in the guidance and Annotation K1)—and the first step in the revised algorithm—specifies that kidney disease be diagnosed on the basis of an eGFR of less than 60 mL/minute or the presence of proteinuria. The second step and annotation K3 in the revised algorithm specify that if evidence of renal toxicity or kidney disease was present while the patient resided at Camp Lejeune or shortly thereafter, and is probably not due to known causes such as diabetes and hypertension, CKD should be assumed to be due to contaminated drinking water exposure.

In the original algorithm, clinicians are expected to determine the cause of CKD on the basis of "history, physical examination, laboratory testing, and imaging tests." In some cases a kidney biopsy may be indicated, such as for nephrotic range proteinuria in the absence of diabetes. However, the committee notes that a kidney biopsy should only be performed when medically indicated for the care of the patient and although it may inform adjudication decisions, it should never be performed solely for the purpose of determining whether the patient should be accepted to the Camp Lejeune program (see revised annotation K1 for algorithm K, Figure 2-2).

The third and fourth steps in the original algorithm K ask the clinician to rule out common causes of CKD (hypertension or diabetes) or other causes (such as volume depletion, severe heart failure, urinary tract obstruction, acute tubular necrosis occurring in the setting of hypotension or nephrotoxic agents, or acute interstitial nephritis often due to drugs) that differ from the text in the guidance (see Table 2-1). Furthermore, the annotations that accompany algorithm K provide much more detail on the clinical signs that may be indicative of other possible causes for CKD than does the text in the guidance. The committee finds that the information presented in the guidance and information presented in the original algorithm K are not parallel.

Similar to previous recommendations made by the NRC (2009), this committee concludes that patients with CKD should have a thorough evaluation. If the evaluation shows that the patient's kidney disease is compatible with another etiology such as diabetic nephropathy or hypertensive nephrosclerosis, the conclusion should be reached that solvent exposure at Camp Lejeune was not the causative agent. If the evaluation does not suggest another etiology, or if the clinical course is atypical for the identified etiology, the patient should be given the benefit of doubt and the conclusion reached that a toxicant exposure at Camp Lejeune may have played a role in the development of CKD. Therefore, the committee finds that the VA's general approach to the guidance and algorithm regarding renal toxicity is appropriate.

Neither the guidance nor the original algorithm K includes other indicators of acute renal injury. Abnormal urinalysis results, serum creatinine, or blood urea nitrogen around the time of exposure and documented in medical records may help a clinician establish that acute effects occurred at or around the time of exposure that later resulted or contributed to CKD. The committee finds that these types of tests, conducted while a patient was in residence at Camp Lejeune, should be considered when determining whether the patient's CKD is related to exposure to contaminated drinking water while at Camp Lejeune. The differences between the guidance text and algorithm K may lead to some confusion and inconsistent conclusions about whether or not a patient's CKD is related to his or her time at Camp Lejeune.

Therefore, the committee recommends that VA consider modifying the guidance and algorithm K—as suggested in revised algorithm K—to indicate that patients presenting with defined reductions in GFR or proteinuria AND who had abnormal renal function tests or urinalysis of unknown etiology while residing at Camp Lejeune should be accepted to the program. The committee also recommends that VA consider accepting into the Camp Lejeune program patients with chronic kidney disease, but without evidence of kidney damage during or around the time of residence at Camp Lejeune, if there are no other more likely causes of their kidney disease.

REFERENCES

Altuntas, G., and Kharasch, E. D. 2002. Biotransformation of l-cysteine S-conjugates and N-acetyl-l-cysteine S-conjugates of the sevoflurane degradation product fluoromethyl-2,2-difluoro-1-(trifluoromethyl)vinyl ether (compound A) in human kidney in vitro: Interindividual variability in N-acetylation, N-deacetylation, and β-lyase-catalyzed metabolism. *Drug Metabolism and Disposition* 30:148-154.

American Kidney Fund. 2012. *Chronic Kidney Disease*. Available at www.kidneyfund.org/kidney-health/kidney-problems/chronic-kidney-disease.html#.VEkCa8lNfSI (accessed December 31, 2014).

Bove, F. J., P. Z. Ruckart, M. Maslia, and T. C. Larson. 2014a. Evaluation of mortality among marines and Navy personnel exposed to contaminated drinking water at USMC Base Camp Lejeune: A retrospective cohort study. *Environmental Health* 13(1):10.

Bove, F. J., P. Z. Ruckart, M. Maslia, and T. C. Larson. 2014b. Mortality study of civilian employees exposed to contaminated drinking water at USMC Base Camp Lejeune: A retrospective cohort study. *Environmental Health* 13:68.

Calvert, G. M., A. M. Ruder, and M. R. Petersen. 2011. Mortality and end-stage renal disease incidence among dry cleaning workers. *Occupational and Environmental Medicine* 68(10):709-716.

Chakrabarti, S. K., and B. Tuchweber. 1988. Studies of acute nephrotoxic potential of trichloroethylene in Fischer 344 rats. *Journal of Toxicological and Environmental Health* 23(2):147-158.

Chawla, L. S., P. W. Eggers, R. A. Star, and P. L. Kimmel. 2014. Acute kidney injury and chronic kidney disease as interconnected syndromes. *New England Journal of Medicine* 371(1):58-66.

EPA (U.S. Environmental Protection Agency). 2011. *Toxicological review of tricholorethylene (CAS No. 79-01-6) in support of summary information on the Integrated Risk Information System (IRIS)*. EPA/635/R-08/011F. U.S. Environmental Protection Agency.

Green, T., J. Dow, C. N. Ong, V. Ng, H. Y. Ong, Z. X. Zhuang, X. F. Yang, and L. Bloemen. 2004. Biological monitoring of kidney function among workers occupationally exposed to trichloroethylene. *Occupational and Environmental Medicine* 61(4):312-317.

IOM (Institute of Medicine). 2003. *Gulf War and health, Volume 2: Insecticides and solvents*. Washington, DC: The National Academies Press.

Irving, R. M., M. E. Pinkerton, and A. A. Elfarra. 2013. Characterization of the chemical reactivity and nephrotoxicity of N-acetyl-S-(1,2-dichlorovinyl)-L-cysteine sulfoxide, a potential reactive metabolite of trichloroethylene. *Toxicology and Applied Pharmacology* 267(1):1-10.

Juhanson, P., K. Kepp, E. Org, G. Veldre, P. Kelgo, M. Rosenberg, M. Viigimaa, and M. Laan. 2008. N-acetyltransferase 8, a positional candidate for blood pressure and renal regulation: resequencing, association and in silico study. *BMC Medical Genetics* 9:25.

Köttgen, A., C. Pattaro, C. A. Böger, C. Fuchsberger, M. Olden, N. L. Glazer, A. Parsa, X. Gao, Q. Yang, A. V. Smith, J. R. O'Connell, M. Li, H. Schmidt, T. Tanaka, A. Isaacs, S. Ketkar, S.-J. Hwang, A. D. Johnson, A. Dehghan, A. Teumer, G. Paré, E. J. Atkinson, T. Zeller, K. Lohman, M. C. Cornelis, N. M. Probst-Hensch, F. Kronenberg, A. Tönjes, C. Hayward, T. Aspelund, G. Eiriksdottir, L. Launer, T. B. Harris, E. Rapmersaud, B. D. Mitchell, E. Boerwinkle, M. Struchalin, M. Cavalieri, A. Singleton, F. Giallauria, J. Metter, I. de Boer, T. Haritunians, T. Lumley, D. Siscovick, B. M. Psaty, M. C. Zillikens, B. A. Oostra, M. Feitosa, M. Province, D. Levy, M. de Andrade, S. T. Turner, A. Schillert, A. Ziegler, P. S. Wild, R. B. Schnabel, S. Wilde, T. F. Muenzel, T. S. Leak, T. Illig, N. Klopp, C. Meisinger, H. E. Wichmann, W. Koenig, L. Zgaga, T. Zemunik, I. Kolcic, C. Minelli, F. B. Hu, Å. Johansson, W. Igl, G. Zaboli, S. H. Wild, A. F. Wright, H. Campbell, D. Ellinghaus, S. Schreiber, Y. S. Aulchenko, F. Rivadeneira, A. G. Uitterlinden, A. Hofman, M. Imboden, D. Nitsch, A. Brandstätter, B. Kollerits, L. Kedenko, R. Mägi, M. Stumvoll, P. Kovacs, M. Boban, S. Campbell, K. Endlich, H. Völzke, H. K. Kroemer, M. Nauck, U. Völker, O. Polasek, V. Vitart, S. Badola, A. N. Parker, P. M. Ridker, S. L. R. Kardia, S. Blankenberg, Y. Liu, G. C. Curhan, A. Franke, T. Rochat, B. Paulweber, I. Prokopenko, W. Wang, V. Gudnason, A. R. Shuldiner, J. Coresh, R. Schmidt, L. Ferrucci, M. G. Shlipak, C. M. van Duijn, I. Borecki, B. K. Krämer, I. Rudan, U. Gyllensten, J. F. Wilson, J. C. Witteman, P. P. Pramstaller, R. Rettig, N. Hastie, D. I. Chasman, W. H. Kao, I. M. Heid, and C. S. Fox. 2010. Multiple new loci associated with kidney function and chronic kidney disease: The CKDGen consortium. *Nature Genetics* 42(5):376-384.

Lash, L. H., J. C. Parker, and C. S. Scott. 2000. Modes of action of trichloroethylene for kidney tumorigenesis. *Environmental Health Perspectives* 108(Suppl 2):225-240.

NCI (National Cancer Institute). 1976. *Carcinogenesis bioassay of trichloroethylene (CAS No. 79-01-6).* Washington, DC: National Institutes of Health, U.S. Department of Health, Education, and Welfare.

NRC (National Research Council). 2006. *Assessing the human health risks of trichloroethylene: Key scientific issues*. Washington, DC: The National Academies Press.

NRC. 2009. *Contaminated water supplies at Camp Lejeune: Assessing potential health effects*. Washington, DC: The National Academies Press.

NRC. 2010. *Review of the Environmental Protection Agency's draft IRIS assessment of tetrachloroethylene*. Washington, DC: The National Academies Press.

NTP (National Toxicology Program). 1986. *Toxicology and carcinogenesis studies of tetrachloroethylene (perchloroethylene) (CAS No. 127-18-4) in F344/N rats and* $B6C3F_1$ *mice (inhalation studies). Technical Report No. 311. NIH 86-2567.* Research Triangle Park, NC: U.S. Department of Health and Human Services, Public Health Service, National Insitutes of Health.

NTP. 1988. *Carcinogenesis studies of trichloroethylene (CAS No. 70-01-6) in four strains of rats (ACI, August, Marshall, Osborne-Mendel) (gavage studies).* Research Triangle Park, NC: U.S. Department of Health and Human Services, Public Health Service, National Institutes of Health, National Toxicology Program.

NTP. 1990. *Carcinogenesis studies of trichloroethylene (without epichlorohydrin) (CAS No. 79-01-06) in F344/N rats and B6C3F1 mice (gavage studies).* Research Triangle Park, NC: U.S. Department of Health and Human Services, Public Health Service, National Institutes of Health, National Toxicology Program.

OMIM (Online Mendelian Inheritance in Man). 2014. *601107, ATP-binding cassette, subfamily C, member 2; ABCC2*. Available at omim.org/entry/601107 (accessed December 13, 2014).

Porro, A., C. Lomonte, P. Coratelli, G. Passavanti, G. M. Ferri, and G. Assennato. 1992. Chronic glomerulonephritis and exposure to solvents: A case-referent study. *British Journal of Industrial Medicine* 49(10):738-742.

Radican, L., D. Wartenberg, G. G. Rhoads, D. Schneider, R. Wedeen, P. Stewart, and A. Blair. 2006. A retrospective occupational cohort study of end-stage renal disease in aircraft workers exposed to trichloroethylene and other hydrocarbons. *Journal of Occupational and Environmental Medicine* 48(1):1-12.

Steenland, K., and S. Palu. 1999. Cohort mortality study of 57,000 painters and other union members: A 15 year update. *Occupational and Environmental Medicine* 56(5):315-321.

Steenland, N. K., M. J. Thun, C. W. Ferguson, and F. K. Port. 1990. Occupational and other exposures associated with male end-stage renal disease: A case/control study. *American Journal of Public Health* 80(2):153-157.

Tsirulnikov, K., N. Abuladze, M. C. Koag, D. Newman, K. Scholz, G. Bondar, Q. Zhu, N. K. Avliyakulov, W. Dekant, K. Faull, I. Kurtz, and A. Pushkin. 2010. Transport of N-acetyl-S-(1,2-dichlorovinyl)-L-cysteine, a metabolite of trichloroethylene, by mouse multidrug resistance associated protein 2 (Mrp2). *Toxicology and Applied Pharmacology* 244(2):218-225.

Veiga-da-Cunha, M., D. Tyteca, V. Stroobant, P. J. Courtoy, F. R. Opperdoes, and E. Van Schaftingen. 2010. Molecular identification of NAT8 as the enzyme that acetylates cysteine S-conjugates to mercapturic acids. *Journal of Biological Chemistry* 285(24):18888-18898.

Vermeulen, R., L. Zhang, A. Spierenburg, X. Tang, J. V. Bonventre, B. Reiss, M. Shen, M. T. Smith, C. Qiu, Y. Ge, Z. Ji, J. Xiong, J. He, Z. Hao, S. Liu, Y. Xie, F. Yue, W. Guo, M. Purdue, L. E. Beane Freeman, V. Sabbisetti, L. Li, H. Huang, N. Rothman, and Q. Lan. 2012. Elevated urinary levels of kidney injury molecule-1 among Chinese factory workers exposed to trichloroethylene. *Carcinogenesis* 33(8):1538-15441.

Walker, K., G. Ginsberg, D. Hattis, D. O. Johns, K. Z. Guyton, and B. Sonawane. 2009. Genetic polymorphism in N-Acetyltransferase (NAT): Population distribution of NAT1 and NAT2 activity. *Journal of Toxicology and Environmental Health. Part B, Critical Reviews* 12(5-6):440-472.

Yamazaki, H., and M. Shimizu. 2013. Survey of variants of human flavin-containing monooxygenase 3 (FMO3) and their drug oxidation activities. *Biochemistry and Pharmacology* 85(11):1588-1593.

Yu, J. F., J. Leng, T. Shen, C. F. Zhou, H. Xu, T. Jiang, S. H. Xu, and Q. X. Zhu. 2012. Possible role of complement activation in renal impairment in trichloroethylene-sensitized guinea pigs. *Toxicology* 302(2-3)172-178.

3

Characterization of Neurobehavioral Effects

The 2009 National Research Council (NRC) report on contaminated drinking water at Camp Lejeune found that there was limited/suggestive evidence of an association between exposure to mixed solvents and neurobehavioral effects. This conclusion was based on the 2003 Institute of Medicine (IOM) report that had similarly found that there was limited/suggestive evidence of an association between "solvents and neurobehavioral effects (that is, abnormal results on neurobehavioral test batteries and symptom findings)." However, the term "neurobehavioral effects" was used in the NRC report (2009) to include such neurobehavioral symptoms as fatigue, lack of coordination, sensory disturbances, confusion, depression, tension, trouble concentrating, and headache; alterations in neurobehavioral tests that indicate deficits in attention, reaction time, visuomotor coordination, motor function, digit symbol, and contrast sensitivity; and certain neuropsychological disorders such as learning or behavioral disorders. That report separated neurologic diseases, such as Alzheimer's disease and Parkinson's disease, from neurobehavioral effects, which left unanswered what those effects indicated in terms of diagnostic entities. Because of the lack of diagnostic specificity, this committee chose to broadly define neurobehavioral effects to include all neurologic and behavioral effects (diseases, disorders, symptoms and deficits) because neurobehavioral symptoms or test findings can be indicative of neurologic or behavioral problems. The 1996 IOM report *Veterans and Agent Orange* stated:

> The central nervous system (CNS) includes the brain and spinal cord, and CNS dysfunction can be subdivided into two general categories, neurobehavioral and motor/sensory. Neurobehavioral difficulties involve two primary categories: cognitive decline, including memory problems and dementia; and neuropsychiatric disorders, including neurasthenia (a collection of symptoms including difficulty concentrating, headache, insomnia, and fatigue), depression, posttraumatic stress disorder (PTSD), and suicide. Other CNS problems can be associated with motor difficulties, characterized by problems such as weakness, tremors, involuntary movements, incoordination, and gait/walking abnormalities.

NEUROBEHAVIORAL AND RELATED EFFECTS

To better define the potential long-term neurobehavioral effects in the U.S. Department of Veterans Affairs (VA) guidance that are associated with an exposure to solvents at Camp Lejeune, this committee reviewed the evidence gathered and synthesized by the 2009 NRC and the 2003 IOM committees and also identified new evidence. Current literature defined the neurobehavioral effects discussed in this chapter; outcomes from the 2009

NRC report without new epidemiologic evidence were not revisited in this report. Literature searches identified new studies published since 2008 in which neurologic and behavioral (including psychological) outcomes were associated with exposure to solvent mixtures, trichloroethylene (TCE), or perchloroethylene (PCE). Those new studies are discussed in conjunction with the evidence proposed in the NRC 2009 report, then synthesized by the committee to determine what neurobehavioral effects might result from exposure to contaminated water at Camp Lejeune. The committee specifically identified neurobehavioral effects solely on the basis of the available literature, using statistically significant findings, the weight of evidence, and the strengths and weaknesses within each key study to determine whether an identified condition should be eligible for coverage under the law. Effects seen in animal studies published since 2008 were not specifically reviewed because the goal of the committee was to identify clinical outcomes in humans.

2009 NRC Report

The NRC (2009) reviewed the scientific evidence on associations between prenatal, childhood, and adult exposures to contaminated water at Camp Lejeune and adverse health effects. Data on accidental and controlled human inhalation and oral exposures and on experimental animal exposures were available and formed the basis for the conclusions in that report.

Neurologic Symptoms, Motor Function, and Sensory Deficits

The NRC report found that most human studies indicate effects on visuomotor and motor function, fatigue, headache, and deficits in concentration, primarily resulting from acute exposures to solvents. Acute inhalation and oral exposure to PCE can induce symptoms of CNS depression (dizziness and drowsiness), electroencephalograpic changes, and neurobehavioral changes such as alterations in flash-evoked visual potentials, deficits in vigilance, and deficits in eye–hand coordination. The effects of long-term occupational exposure to TCE include memory loss, mood swings, the impairment of cognitive function, and olfactory and trigeminal neuropathy (NRC, 2009). Oral doses of PCE given as an anthelminthic (de-wormer) resulted in narcotic effects and various associated changes, such as inebriation, perceptual distortion, and exhilaration (ATSDR, 1997). The NRC report also cited a study (Reif et al., 2003) that evaluated neurobehavioral function in 184 adults who had been exposed to TCE-contaminated drinking water many years before testing. Higher exposures were associated with poorer performance on several tests (such as digit symbol substitution test, contrast sensitivity C test, and contrast sensitivity D test) and with increased neurobehavioral symptoms (such as confusion, depression, and tension).

Amyotrophic Lateral Sclerosis, Alzheimer's Disease, and Multiple Sclerosis

In the NRC report, several neurologic diseases and endpoints were assessed to determine if they were associated with exposure to TCE, PCE, or solvent mixtures. Specifically, amyotrophic lateral sclerosis (ALS), Parkinson's disease, multiple sclerosis (MS), and Alzheimer's disease were considered. The report concluded that there was inadequate/insufficient evidence to determine whether an association exists between exposure to solvents and ALS, MS, or Alzheimer's disease (NRC, 2009).

Parkinson's Disease

The NRC report concluded that there was inadequate/insufficient evidence to determine whether an association exists between solvent exposure and Parkinson's disease (NRC, 2009). Two case-control studies (Dick et al., 2007; McDonnell et al., 2003) and one occupational cohort (Gash et al., 2008) were evaluated by the NRC committee to assess the relationship between exposure to solvents and Parkinson's disease. The first case-control study showed a trend of increasing odds of developing the disease with increasing duration of occupational exposure; however, the study did not account for other possible risk factors or confounders. The second study did not show any association between solvent exposure and the risk of developing Parkinson's disease, but this study relied on individual recall regarding occupational and hobby-related exposures to solvents. The final study was an occupational cluster investigation that showed that three workers diagnosed with Parkinson's disease had workstations adjacent to a TCE source.

Updated Literature

The committee identified several new epidemiologic studies that looked at the association between exposure to solvents, particularly TCE and PCE, and neurobehavioral effects such as motor function as well as neurobehavioral symptoms resulting from neurological diseases such as Parkinson's disease. The committee did not assess new toxicological studies on these solvents because they currently would not have clinical applications. However, one paper (Bale et al., 2011) reviewed mechanistic studies using TCE, PCE, or dichloromethane and proposed mechanisms of action for the different neurological effects observed in those studies. The authors concluded that cognitive decrements may be due to changes in cholinergic transmission, while visual system changes were mediated by N-methyl-D-aspartic acid (NMDA)-glutamate or the nicotinic acetylcholine receptor pathway. The disruption of sodium channel function may lead to demyelination associated with multiple sclerosis. Data for these solvents were insufficient to propose a mechanism for ototoxicity or sleep-cycle changes.

Neurologic Symptoms, Motor Function, and Sensory Deficits

This committee's updated literature search identified one new study that addressed neurologic effects resulting from solvent exposure. Static postural sway and hand tremor parameters were evaluated in 57 workers occupationally exposed to TCE for 0.1 to 37 years (mean 10.9 years); 60 unexposed workers served as controls (Murata et al., 2010). A cumulative exposure index was calculated by multiplying total urinary trichloro-compound levels by work duration. Neuromotor function tests were conducted on a Friday after the work shift. Maximum ambient TCE was estimated at less than 22 ppm, but air measurements were not taken. Sway area, transversal sways, and sagittal sways with eyes open were all significantly greater in the exposed workers than in the controls ($p = < 0.001$, 0.012, and 0.029, respectively). Hand tremor intensities in the dominant hand were significantly larger in exposed workers than in the controls ($p = 0.038$), but there was no significant difference for the non-dominant hand. A trend of greater sway and increased tremor intensity was seen with increasing exposure (as measured by a cumulative exposure index), but the effect was not statistically significant, probably because of the small number of individuals in each exposure category.

This committee concluded, based on the 2009 NRC report and the updated literature, that the best characterized neurologic effects associated with solvent exposure, in particular exposure to TCE and PCE, were deficits in visuomotor function, motor function, memory, and concentration. Based on the evidence, the committee is interpreting "deficits in concentration" to mean attentional disorders.

Amyotrophic Lateral Sclerosis, Alzheimer's Disease, and Multiple Sclerosis

In the 2009 NRC report, several neurologic diseases and endpoints were assessed to determine if they were associated with exposure to TCE, PCE, and solvent mixtures. Specifically, the report considered ALS, Parkinson's disease, MS, and Alzheimer's disease. This committee considers each of these diseases to have neurobehavioral effects that could lead to their diagnosis and therefore looked at any new literature since the 2009 report on exposure to solvents and the development of these neurologic diseases.

No new evidence provided additional support for a relationship between exposure to solvents and the development of ALS, Alzheimer's disease, or MS. However, two investigations of Camp Lejeune military personnel and civilians compared mortality from ALS and MS with that of military personnel and civilians stationed at U.S. Marine Corps Camp Pendleton in California.

In two retrospective cohort studies (Bove et al., 2014a,b), both civilian employees and military personnel stationed at Camp Lejeune were evaluated for exposure to contaminated drinking water and risk of mortality from cancers and other chronic diseases. It should be noted that only mortality resulting from a disease was examined, not the development or prevalence of the disease itself. Both populations were matched to control cohorts from Camp Pendleton. The exposures of Camp Lejeune employees and military personnel were estimated using average monthly contaminant concentrations in the drinking water during the period of their employment or residence on base. All cohorts were identified through the Defense Manpower Data Center files, with vital status at follow up obtained through the Social Security Administration Death Master File and the National Death Index.

Among military personnel, there were 27 and 21 deaths from ALS at Camp Lejeune and Camp Pendleton, respectively, and 10 and 12 deaths from MS at the two camps. These results are elevated compared to the general population but do not reach statistical significance (Bove et al., 2014a). Among civilians who worked at Camp Lejeune or at Camp Pendleton there were one and four deaths from ALS, respectively, and there was one death from MS at each camp. Because there were so few cases, the authors could not compute a hazard ratio comparing the two camps for either disease (Bove et al., 2014b). Thus, no increase in mortality from ALS or MS was observed in these study populations.

Parkinson's Disease

The committee identified four new studies that address solvent exposure and Parkinson's disease that have been published since the NRC report.

In a case-control study to determine the association between occupational exposure to specific solvents—including TCE and PCE—and the development of Parkinson's disease, 99 all-male twin pairs discordant for Parkinson's disease were identified from the National Academy of Sciences/NRC World War II Veteran Twins Registry (Goldman et al., 2012). Occupational solvent exposure was assessed through a questionnaire. Exposure to TCE was associated with a significantly increased risk of Parkinson's disease (OR [odds ratio] = 6.1, 95% CI [confidence interval] 1.2–33) and tended toward significance for PCE (OR = 10.5, 95% CI 0.97–113). The risk was also significantly increased for the combined variable of TCE or PCE exposure (OR = 8.9, 95% CI 1.7–47). In 48% of the pairs, at least one twin was exposed to one or more of the six solvents studied. The mean duration of exposure to TCE or PCE was 9.0 years in the control twin compared with 18.5 years in the twin diagnosed with Parkinson's disease ($p = 0.009$), suggesting that the duration of exposure was a factor in the development of the disease.

In the retrospective cohort studies of mortality in the Camp Lejeune and Camp Pendleton populations, the civilian cohort included 4,647 full-time civilian employees with a median age of 58 years who were employed at Camp Lejeune during 1973–1985 (Bove et al., 2014b). Controls were 4,690 full-time civilian employees (median age 60 years) at Camp Pendleton. The standardized mortality ratio (SMR) for Parkinson's disease in the Camp Lejeune cohort was 2.91 (95% CI 0.71–5.11) versus an SMR of 0.88 (95% CI 0.24–2.26) for the Camp Pendleton cohort. There was a nonsignificant increase in the risk of mortality from Parkinson's disease among the Camp Lejeune cohort (HR = 3.13, 95% CI 0.76–12.86) compared with the Camp Pendleton cohort, adjusted for sex, race, occupation, and education. There were five cases of Parkinson's disease in the Camp Lejeune cohort, as compared with four cases from Camp Pendleton. Four of the five Camp Lejeune cases were associated with a cumulative exposure above the median for TCE and PCE as well as for vinyl chloride and benzene resulting in hazard ratios of greater than 2.50 ($p \leq 0.05$).

The military cohort included 154,932 marine and Navy personnel with a median age of 49 years who began active-duty service between April 1975 and December 1985, and who were stationed at Camp Lejeune at some time during that period. The controls were 154,969 personnel stationed at Camp Pendleton at any time during the same active-duty interval. SMRs were not calculated for Parkinson's disease because there were fewer than five cases in each cohort. The committee notes that the military Camp Lejeune cohort is still too young to be informative about the risk of death from Parkinson's disease. At the end of the 2008 mortality follow up, the median age was 49, and only 2.7% of cohort members were 55 or older (Bove et al., 2014a). Because Parkinson's disease is a rare condition before age 50 (National Parkinson Foundation, 2014) and mortality occurs several years after diagnosis, a longer follow up is needed to provide meaningful results on this disease.

In a recent review, Lock et al. (2013) concluded that neither toxicologic nor epidemiologic studies present clear evidence that any specific solvent or class of solvents is an established cause of Parkinson's disease. However, based on the findings of Goldman et al. (2012) and Bove et al. (2014a,b), as well as limited data from the NRC

(2009) and the U.S. Environmental Protection Agency (EPA) (2011) assessments, the committee finds that TCE and similar solvents may have potential etiologic relevance in the development of Parkinson's disease.

The committee concludes that Parkinson's disease is a neurobehavioral effect that may have resulted from the consumption of the contaminated drinking water at Camp Lejeune. This conclusion is based on the positive trends of increased risks from occupational and drinking water exposures reported by Goldman et al. (2012), NRC (2009), and Bove et al. (2014b). Despite the limitations of these studies, such as lack of statistical significance, the potential for recall bias, and the lack of incidence data pertaining to Parkinson's disease, the committee recommends including Parkinson's disease as an outcome associated with exposure to TCE and PCE. Because of the slow onset of Parkinson's disease, patients developing it years after their exposure, regardless of their age at exposure, may have not had symptoms at the time of exposure. Patients who have Parkinson's disease now or who develop it in the future and are otherwise eligible for the Camp Lejeune program should be covered by the guidance for neurobehavioral effects even if the symptoms were not apparent during their time at Camp Lejeune.

The committee recommends that VA consider adding Parkinson's disease in the clinical guidance and in algorithm B as a neurobehavioral effect that may result from exposure to contaminated drinking water at Camp Lejeune.

IN UTERO AND CHILDHOOD EXPOSURES

Although the majority of the diseases listed in the Janey Ensminger Act have been associated with adult exposure to solvents, the act also acknowledges that pregnant women who resided at Camp Lejeune may have ingested the contaminated drinking water and by doing so inadvertently exposed their fetuses. In addition, children living on Camp Lejeune also consumed the contaminated water. The committee believes that the health impacts of the consumption of solvent-contaminated drinking water on fetuses, infants, and children need to be considered in the VA guidance (discussed later) and therefore, the effects of solvents on children, who were exposed in utero or during early childhood—and who are now adults—are discussed in this section. Furthermore, because of the neurologic involvement and behavioral effects resulting from some types of birth defects, such as neural tube defects, the committee also reviewed the evidence for these types of outcomes.

2009 NRC REPORT

The 2009 NRC report describes how children are particularly susceptible to contaminants such as solvents by noting the following:

> [T]here were "windows of vulnerability" or short periods of early human development when chemical exposures may significantly alter organ function or structure. Potentially vulnerable targets in infants and young children include the endocrine, reproductive, immune, visual, and nervous systems. Little information is available on the effects of TCE, PCE, and other solvents on the development of those organ systems in laboratory animals or humans.

The NRC report found that only a few studies assessed neurotoxic outcomes in rats resulting from the exposure of the fetus to low concentrations of TCE in drinking water during pregnancy and lactation. In all studies, the doses to the animals were below those causing overt maternal or fetal toxicity. Reported neurobehavioral effects included increased activity, reduced 2-deoxyglucose uptake in the brain, learning deficits, and reduced hippocampal myelin. The effects of PCE exposure during development included the neurobehavioral impairment of rats and mice on certain days of testing, reductions in acetylcholine and dopamine, changes in motor activity and attenuation of habituation, and altered pain and seizure thresholds. These studies of behavioral effects in rats and mice exposed to PCE prenatally or postnatally further suggest that there may be sensitive windows for neurobehavioral impairment during development.

There were few epidemiologic data available to characterize the effects of solvents in children exposed in utero or postnatally in the 2009 NRC report. This exposure pathway was not considered in the earlier IOM report that assessed potential health outcomes in veterans of the 1990–1991 Gulf War who were exposed to solvents

(IOM, 2003). That report assumed that there were no pregnant female service members deployed in the 1990–1991 Gulf War. In the NRC report, the few studies that were identified focused on childhood cancers that may have resulted from solvent exposure. Only one study available at that time assessed the effects of prenatal exposure to PCE on learning and behavioral disorders; these exposures came from contaminated drinking water in Cape Cod (Janulewicz et al., 2008).

The only conclusion that the 2009 NRC report made regarding adverse outcomes in children was "that there continues to be inadequate/insufficient evidence to determine whether an association exists between chronic exposure to TCE or PCE and congenital malformations." The congenital malformations assessed included heart defects, neural tube defects, and oral clefts.

Updated Literature

Most of the new literature identified by the committee was the result of a series of epidemiologic studies on a Cape Cod, Massachusetts, population that had been exposed to PCE. From 1968 through 1980, PCE had leached into the drinking water supplies from lined pipes installed in the public water distribution systems of several towns on Cape Cod. Aschengrau et al. published a series of population-based retrospective cohort studies examining the association between prenatal and postnatal drinking water exposure to PCE and a number of adverse neurobehavioral outcomes (Aschengrau et al., 2011, 2012; Getz et al., 2012; Janulewicz et al., 2008, 2012, 2013) and congenital anomalies (Aschengrau et al., 2009). Exposure estimates came from the modeled cumulative mass of PCE entering the homes of study participants and were not a direct measure of PCE intake by the subjects. Cumulative exposure during gestation and early childhood was calculated as the sum of 75% of the estimated mass of PCE delivered to the residence during the birth year plus the estimated mass of PCE from the month and year following birth through the month and year of the fifth birthday. Exposure assessments beyond the fifth birthday could not be conducted because of limitations in the water systems records. Exposed and unexposed populations were cross-matched with a database of all street locations served by the contaminated pipes. Because nearly all subjects with prenatal exposure also had early childhood exposure, the impact of prenatal exposure alone could not be determined in these studies.

Other studies of in utero or childhood exposure to solvents include a case control study of mothers from Camp Lejeune during the time of the water contamination (Ruckart et al., 2013), a population-based National Birth Defects Prevention Study (Desrosiers et al., 2012), and two studies of environmental exposures (Storm et al., 2011; Till et al., 2005). In the sections below, the committee considers the neurobehavioral effects that have been reported in the children exposed to solvents in utero or in childhood at Cape Cod, Camp Lejeune, and elsewhere.

Birth Defects Affecting the Nervous System

A case-control study was conducted to determine whether children born to mothers with residential exposure to contaminated drinking water at Camp Lejeune during pregnancy were more likely to have childhood hematopoietic cancers, neural tube defects, or oral clefts (Ruckart et al., 2013). The exposed population included live births between 1968 and 1985 to mothers who resided on base at any time during their pregnancy. Parents of 12,598 children were asked during a telephone interview if their child had a birth defect. The risk for neural tube defects associated with average first-trimester exposures were increased nonsignificantly for TCE above 5 parts per billion (ppb) (OR = 2.4, 95% CI 0.6–9.6). A monotonic exposure response relationship was observed in those exposed to less than 5 ppb (OR = 1.1, 95% CI 0.3–3.5) and greater than 5 ppb (OR = 2.4, 95% CI 0.6–9.6) compared with those who were unexposed. (Five ppb is the maximum contaminant level; see Chapter 1 for a description.)

A Cape Cod population cohort of 1,658 children exposed prenatally to PCE and 2,999 unexposed children was also examined for risks of congenital anomalies (Aschengrau et al., 2009). No meaningful increases in ORs were seen for cardiac and musculoskeletal malformations, and there were too few exposed cases to estimate ORs for eye; ear, face, and neck; respiratory; and other anomalies. Among children with any prenatal exposure, there was a nonsignificant increase in the ORs for neural tube defects (OR = 3.5, 95% CI 0.8–14.0). The neural tube defects observed in the cohort included four cases of anencephaly among exposed children versus none in the unexposed children, one case of spina bifida among the exposed children versus three cases among unexposed children, and one case of Arnold-Chiari malformation among exposed versus none among unexposed children.

Risks of neural tube defects from a maternal occupational exposure to organic solvents were assessed using data from the population-based National Birth Defects Prevention Study (Desrosiers et al., 2012). The maternal occupational exposure period was restricted to 1 month prior to the estimated date of conception through the end of the first trimester; jobs were coded by occupation and industry and assessed for exposure to 10 organic solvents, including TCE and PCE. Regression analyses were used to determine associations between solvent class (chlorinated, Stoddard, aromatic) and outcome. A total of 511 neural tube defect cases with 2,972 controls were included in the analyses; exposure to chlorinated solvents was associated with a statistically significant increased risk of neural tube defects (OR = 1.96, 95% CI 1.34–2.87).

Risks of congenital anomalies in children born to mothers exposed to TCE, PCE, or other solvents during pregnancy have been evaluated in several studies. An association between neural tube defects and drinking water exposure (Aschengrau et al., 2009; Ruckart et al., 2013) or occupational exposure (Desrosiers et al., 2012) has been shown. The committee concludes that neural tube defects may have resulted from in utero exposures to these solvents in the contaminated drinking water at Camp Lejeune.

The committee recommends that VA consider including neurobehavioral effects as a result of neural tube defects to the Camp Lejeune clinical guidance and in algorithm B-1.

Neuropsychological Performance

Performance on a battery of neuropsychological tests was assessed in a small cohort of 35 adults who had been exposed to PCE at Cape Cod in utero or during early childhood between 1969 and 1983 and also in 28 unexposed subjects (Janulewicz et al., 2012). No associations were found between prenatal and early postnatal exposure to PCE and decrements on tests that assess abilities in the domains of omnibus intelligence, academic achievement, and language. Trends were found among exposed individuals both for mood alterations and for slightly worse performances in various domains, including visuospatial, learning and memory, attention, fine motor speed, and executive function, but the effects were not statistically significant, most likely because of the small sample size.

A sample of 1,349 exposed and 737 unexposed children was evaluated for risk of learning and behavioral disorders following prenatal and early postnatal exposure to PCE (Janulewicz et al., 2008). All the children were born between 1969 and 1983 to mothers who lived on Cape Cod at the time of birth; enrollment occurred during 2002–2003. The measures of learning, attention, and behavior used in the study included whether the child ever had a diagnosis of attention deficit disorder or hyperactive disorder, ever received tutoring for reading or math, ever had special class placement for academic or behavioral problems, ever had an individual education plan, or ever repeated a school grade, as well as the child's highest level of education achieved. No associations were found between exposures and the maternal reports of any measured outcome.

Psychiatric Disorders

Another cohort of the Cape Cod population with 831 exposed and 547 unexposed children enrolled during 2006–2008 was evaluated for an affinity for risky behaviors and for the occurrence of mental illness. Risky behaviors included in the study were smoking, drinking, or illicit drug use as a teen or adult (Aschengrau et al., 2011). Individuals with any level of exposure during gestation and early childhood were more likely than unexposed subjects to have used two or more major illicit drugs as a teenager (risk ratio [RR] = 1.4, 95% CI 1.0–1.8) or as an adult (RR = 1.3, 95% CI 1.0–1.6). Individuals in the highest tertile of exposure (that is, greater than the 67th percentile) during gestation and early childhood were 50 to 60% more likely to have used two or more major illicit drugs as a teenager (RR = 1.6, 95% CI 1.2–2.2) or as an adult (RR = 1.5, 95% CI 1.2–1.9). The specific drugs for which increased risks were observed included cocaine and crack cocaine, psychedelics and hallucinogens, and club and designer drugs, Ritalin without a prescription, and heroin (RRs = 1.4–2.1). Increases in the risk of certain drinking behaviors were seen only among highly exposed subjects, with no evidence of a dose–response relationship. For example, only individuals in the highest tertile of exposure during gestation and early childhood experienced increases in the risk of drinking more than 8 days/month as a teenager (RR = 1.6, 95% CI 1.1–2.3).

Mental illnesses (depression, bipolar disorder, PTSD, and schizophrenia) were also assessed in the Cape Cod cohort (Aschengrau et al., 2012). Subjects with any exposure during gestation and childhood were 1.8 times more likely to have developed bipolar disorder (95% CI 0.9–3.5), although the effect was not statistically significant. However, when the analysis was restricted to subjects in the highest tertile of exposure, a significantly increased risk for bipolar disorder was observed (RR = 2.7, 95% CI 1.3–5.6). The risk of PTSD was greater for subjects with any exposure during gestation and childhood (RR = 1.5, 95% CI 0.9–2.5), but the effect was not statistically significant. While there were too few cases of schizophrenia to examine a dose–response relationship, three of the four schizophrenia cases were in the exposed group (RR = 2.1, 95% CI 0.2–20.0 for any exposure). No associations were found between PCE exposure and an increased risk of depression among exposed subjects.

The available studies of the effects of solvents in children and adults who were exposed in utero or in childhood in Cape Cod found a number of neurobehavioral effects. The limitations of these studies include their retrospective nature, modeled exposure estimates, and self-reported mental illnesses. However, the major strengths of the studies were that exposed and unexposed groups were from the same geographic location, population characteristics were similar between the two groups, and similar proportions of participants and nonparticipants were exposed to PCE, which reduced selection bias. No other studies were identified that examined these psychological and psychosocial outcomes in association with in utero or childhood exposure to PCE, TCE, or other solvents. Thus, although the positive findings reported for the Cape Cod cohorts for illicit drug use and bipolar disorder associated with in utero and early childhood exposures to PCE have not been confirmed by research in other populations, the committee agreed that the studies provide important information on such exposures and warrant further research.

Committee members were not in agreement on whether the two studies on illicit drug use and bipolar disorder (Aschengrau et al., 2011, 2012) provided enough evidence to warrant a recommendation on the inclusion of these two neurobehavioral effects in the guidance and in the algorithms.

Nevertheless, in keeping with the VA policy that "in cases where there is reasonable doubt as to the diagnosis or primary cause for the diagnosis, clinicians should resolve in favor of the Camp Lejeune veteran or family member," the committee recommends that VA consider including adolescent and adult illicit drug use and bipolar disorder as neurobehavioral effects in the Camp Lejeune clinical guidance and in algorithm B-1.

ADDITIONAL ENDPOINTS

The literature identifies a number of endpoints of concern, other than those identified in the preceding sections, but the evidence precluded the IOM committee from making a recommendation on the findings at this time. Trends for problem drinking and alcoholism, PTSD, and schizophrenia were found in some populations (Aschengrau et al., 2009, 2011, 2012). However, the strength of these results was limited by the small numbers of cases observed in the study populations, a lack of dose–response effects, and failures to reach statistical significance.

Other endpoints of concern are discussed in the following sections.

Structural Brain Changes

The final neurobehavioral-related outcome examined in a cohort from Cape Cod was overt structural brain changes as detected with structural magnetic resonance imaging (MRI) (Janulewicz et al., 2013). Brain imaging was performed on 26 exposed and 16 unexposed subjects in order to obtain measurements of specific brain regions. No statistically significant differences were found between exposed and unexposed subjects on the measures of white matter hypointensities, white matter volumes, or gray matter volumes.

Vision

The 2003 IOM and the 2009 NRC reports found that there was inadequate/insufficient evidence to determine whether an association existed between exposures to solvents and long-term reduction in color discrimination. Peer consultants for the 2004 EPA draft *Summary Report of the Peer Review Workshop on the Neurotoxicity of Tetrachloroethylene (Perchloroethylene) Discussion Paper* suggested that:

> contrast sensitivity loss may reflect impaired function throughout the brain, because contrast sensitivity is affected by retinal, optic nerve, or central brain dysfunction (EPA, 2004). Nonetheless, drawing strong conclusions from these studies is difficult, particularly in light of the paucity of data on this test in occupational populations with higher exposure concentrations and in animal studies. (EPA, 2012)

The committee identified three additional studies of vision effects resulting from PCE exposure (two of which included children) and one study of the effects of organic solvent exposure on vision. In the first study, deficits in color vision and contrast sensitivity were assessed in a small cohort of 29 exposed and 25 unexposed Cape Cod adults who were about 30 years of age at testing, and all of whom had been exposed to PCE during the prenatal and early postnatal period (Getz et al., 2012). None of the participants had subjective visual complaints. However, the participants in the higher PCE exposure group exhibited lower contrast sensitivity at intermediate and high spatial frequencies than did the unexposed participants, although the differences were generally not statistically significant. The exposed participants also exhibited poorer color discrimination than the unexposed participants; the mean difference in the Farnsworth color confusion index between PCE-exposed and unexposed participants was 0.05 (95% CI 0.003–0.10; $p = 0.04$).

Storm et al. (2011) evaluated the effects of current exposure to PCE on visual contrast sensitivity in 54 adults and 50 children residing in buildings co-located with a dry cleaner using PCE. Increases in PCE levels in indoor air, breath, and blood were significantly ($p = 0.02$) associated with decreased visual contrast sensitivity in the children's worse-performing eyes at the specific spatial frequency of 12 cycles per degree. Adult visual contrast sensitivity was not affected.

Till et al. (2005) examined visual abnormalities in 21 infants (mean age 12 months) exposed prenatally to organic solvents and compared them with 27 unexposed age-matched infants; the specific chemicals were not identified. Exposed infants showed a significant reduction in contrast sensitivity ($p < 0.001$) as well as abnormal visual evoked potential responses to the red–green onset stimulus ($p < 0.01$) but not to either blue–yellow or achromatic stimuli.

Vision abnormalities, including diffuse color vision losses ($p < 0.01$), have been reported for 25 adults occupationally exposed (gas station workers) to mixed solvents (Costa et al., 2012). Gobba and Cavalleri (2000)[1] found that in 33 PCE-exposed dry cleaners who had occupationally induced color vision loss, no change in color perception was observed for those workers whose exposure decreased, while in others a rise in PCE levels was followed by a significant worsening of color vision.

The committee acknowledges that the visual deficits found in these studies are subclinical, that several studies had very small samples sizes, and that the evidence necessary to assess whether the effects are short term or long term is lacking. Nevertheless, the committee finds that the weight of evidence indicates deficits in contrast sensitivity and color discrimination may result from exposure to TCE, PCE, or solvents and are neurobehavioral effects that may result from prenatal, childhood, and adult exposures to solvents in the contaminated drinking water at Camp Lejeune.

The committee recommends that problems with contrast sensitivity and color discrimination be included in the clinical guidance and in algorithm B as neurobehavioral effects that may result from exposure to contaminated drinking water at Camp Lejeune, although it recognizes that these are typically subclinical (that is, they are not detectable upon routine examination), and no treatments for them are currently available. Given their subclinical nature, the committee further recommends that patients not be screened for these conditions unless there is a clear reason to do so (for example, the patient reports visual problems), and that the results of any screening or testing for visual problems should be noted in the patient's record.

[1] This study is included here despite the date of publication because it was not cited in the NRC report (2009).

DISCUSSION OF THE GUIDANCE AND ALGORITHM

In this section, the committee considers the text of the VA guidance and discusses the accuracy and clarity of the guidance and algorithm B as well as inconsistencies within and between the guidance and algorithm B. The committee also suggests specific language for revising the original algorithm B and proposes a new algorithm B-1 for neurobehavioral effects associated with solvent exposures in utero and in childhood. Algorithm B applies to exposures for all ages.

The guidance currently has a short section for clinicians on what is meant by neurobehavioral effects, which conditions qualify as covered conditions, and what signs or symptoms should be determined to have been present when veterans or family members were exposed to contaminated drinking water during or shortly after their time at Camp Lejeune. The committee notes that the neurobehavioral effects presented in the text (pp. 8–9) are not entirely consistent with those given in algorithm B, nor do they reflect the supporting data on those effects as presented in the 2009 NRC report (see Table 3-1).

For example, in the text of the guidance, deficits in color vision are not mentioned; however, algorithm B specifically lists problems with color vision as one of the few effects that have been identified as a neurobehavioral symptom for the Camp Lejeune program.

In Box 4 of the original algorithm B, "Reductions in color discrimination" is also listed as a diagnosed (and

TABLE 3-1 Neurobehavioral Effects as Given in Algorithm B and the Revised Algorithm B

Algorithm B	Revised Algorithm B
Identified symptoms include:	
Delayed reaction times	Delayed reaction times
	Parkinson's disease (added)
Problems with	Problems with memory
Short-term memory	Problems with visuomotor function
Visual perception	Problems with color discrimination
Color vision	Problems with attention
Attention	Problems with contrast sensitivity (added)
	Problems with motor function (e.g., hand tremor, postural sway) (added)
Are the neurobehavioral symptoms caused by a diagnosed neurologic condition such as:	
Alzheimer's disease or other dementia	Alzheimer's disease or other dementia
Amyotrophic lateral sclerosis	Amyotrophic lateral sclerosis
Multiple sclerosis	Multiple sclerosis
Parkinson's disease (deleted)	
Reductions in color discrimination (deleted)	Genetic color blindness (added)
	Other defined basal ganglia disease such as striatonigral degeneration, multiple system atrophy with orthostatic hypotension (Shy-Drager syndrome), progressive supranuclear palsy (added)
	Cerebrovascular disease (added)
	Primary or metastatic brain tumors not associated with covered cancers (added)
Are the neurobehavioral symptoms caused a diagnosed psychiatric disorder manifest before exposure?	
Bipolar	Bipolar depression
Schizophrenia	Schizophrenia
Posttraumatic stress disorder	Posttraumatic stress disorder
Obsessive compulsive disorder	Obsessive compulsive disorder
Panic disorder	Panic disorder
Attention deficit hyperactivity disorder	Attention deficit hyperactivity disorder
	Unipolar depression (added)

therefore exclusionary) neurologic condition that may cause the neurobehavioral symptoms otherwise associated with residence at Camp Lejeune. This is confusing and should be clarified and made consistent both in the guidance text and the algorithm B.

The guidance states on page 8 that neurobehavioral effects "*would likely have been manifest at the time of exposure or shortly thereafter*" (emphasis in guidance) and further cautions clinicians that neurobehavioral effects that first occur after a long asymptomatic period are "not likely to be secondary to the contaminated water at Camp Lejeune," nor are they likely to have persisted and to require treatment at this time. In essence, the guidance suggests that neurobehavioral effects in adults resulting from exposure to contaminated water at Camp Lejeune are unlikely to require treatment at the current time; impacts on children exposed in utero or in childhood are not discussed. It might be helpful for clinicians if the guidance text clarified that for residents who have neurobehavioral symptoms that have persisted since their time at Camp Lejeune, those symptoms are indeed eligible to be covered; this is clear in the algorithm but not in the text.

The guidance contains a paragraph on page 9 that discusses neurobehavioral effects that are associated with long-term exposure to mixed solvents, occupational exposures, and chronic low-level exposures. One citation is given for each statement. The committee finds that this paragraph may present an incomplete picture of the neurobehavioral effects that have been associated with solvent exposure in the recent literature (e.g., EPA, 2011, 2012). It is unclear what criteria VA used to select the studies mentioned in the paragraph (that is, Chen et al., 2001; Dick et al., 2010; Flodin et al., 1984; and van Valen et al. 2009) as they are not necessarily representative of the epidemiological literature on TCE, PCE, and other solvents.

The classification scheme for neurobehavioral effects is taken from the National Institute for Occupational Safety and Health (NIOSH) bulletin on organic solvent neurotoxicity (NIOSH, 1987). The committee notes that this document is for occupational exposures and that although it includes chronic exposure and solvent abuse situations, these are not representative of the exposures at Camp Lejeune. The neurobehavioral effects discussed in the NIOSH bulletin do not parallel those in the guidance or the algorithm. Should the NIOSH classification scheme be retained in the guidance, the committee suggests that the guidance explain to clinicians how they should use the algorithm when evaluating Camp Lejeune program participants.

Given the inconsistencies between the guidance and algorithm B for neurobehavioral effects in adults following exposure to Camp Lejeune drinking water, the committee recommends that the VA clinical guidance and algorithm B be revised to be consistent and to reflect recent literature.

Finally, the committee notes that although the Janey Ensminger Act specifically states that *family member* includes those who were in utero while at Camp Lejeune, the guidance does not address prenatal exposure and possible subsequent neurobehavioral effects. Based on the evidence considered in the earlier section "In Utero and Childhood Exposures," the committee believes it is important that the guidance address prenatal and childhood exposure to Camp Lejeune contaminants as the outcomes differ and are not captured in the current guidance or in algorithm B.

The committee has determined that it is reasonable to expect that the exposure period relevant for the "prenatal and adolescence exposure algorithm" should encompass the time period from conception through the age of 18. Although the Cape Cod studies modeled exposure only through the age of 5 years because of data limitations, the human brain continues to develop after age 5 through multiple cellular processes (e.g., gliogenesis, synaptic pruning, synaptic remodeling, and apoptosis). This continued brain development has been captured and analyzed using MRI technology with individuals from the ages of 2 weeks though 18 years (Brain Development Cooperative Group, 2012; Sanchez et al., 2012). If the exposure occurred after the age of 18, the individual would be evaluated using the adult exposure algorithm.

Thus, the committee recommends that VA consider including in the clinical guidance a new algorithm B-1 for neurobehavioral effects specific to prenatal and childhood exposure at Camp Lejeune.

Algorithm B: Adult Exposure

Algorithm B is for neurobehavioral symptoms seen in Camp Lejeune residents who were exposed at any age to the contaminated drinking water. The symptoms must have begun while the patient was at Camp Lejeune and have continued through to the present. The committee finds that this accurately reflects what the evidence says about the development of these conditions, with the exception of Parkinson's disease, which may have developed at any time from the point of exposure to the present; the revised algorithm accounts for this difference. The original algorithm and guidance specify that the symptoms must not be associated with more common neurological or psychiatric conditions not known to be related to exposure to contaminated drinking water at Camp Lejeune. Because different clinicians may be diagnosing the conditions in Box 4 (assessed by a neurologist) and Box 5 (assessed by a psychiatrist), it is helpful to differentiate the conditions in each one. Table 3-1 presents the neurobehavioral symptoms and exclusionary conditions presented in algorithm B and in the guidance compared to the committee's suggested changes to that algorithm. A revised algorithm B is presented in Figure 3-1.

Suggested Algorithm B-1 for In Utero and Childhood Exposure

As discussed earlier in this chapter, in utero and childhood exposures were possible at Camp Lejeune. However, the committee finds that if the patient was exposed as a child or fetus the neurobehavioral deficits caused by exposure may not have been manifest at the time of exposure. Therefore, the committee finds that it is prudent to consider including the following neurobehavioral conditions in adults who were exposed in utero or as children (up until age 18 years) to contaminated drinking water at Camp Lejeune, although not all committee members were in agreement on the addition of illicit drug use and bipolar disorder:

- Illicit drug use
- Bipolar disorder
- Neurobehavioral effects caused by neural tube defects

The suggested Algorithm B-1 is presented in Figure 3-2.

ANNOTATIONS FOR REVISED ALGORITHM B:
ADHD = attention deficit hyperactivity disorder; ALS = amyotrophic lateral sclerosis; OCD = obsessive compulsive disorder; PTSD = posttraumatic stress disorder.
B1—Identified neurobehavioral symptoms include delayed reaction times and problems with memory, visuomotor function, attention, motor function (such as tremor), contrast sensitivity, and color discrimination.
B2—Applicant did not have symptoms at the time of exposure, or documented symptoms first occurred a prolonged time after residence at Camp Lejeune ceased. Research to date has not shown any evidence of onset or progression of symptoms after cessation of exposure.
Applicant does not have neurobehavioral symptoms as a covered condition and is not eligible for the Camp Lejeune program at this time.
B3—Applicant has a neurological condition that commonly causes those neurobehavioral symptoms. Other basal ganglion diseases include striatonigral degeneration, multiple system atrophy, orthostatic hypotension, and progressive supranuclear palsy. The 2003 IOM and 2009 NRC reviews found no evidence or inadequate or insufficient evidence of an association between these neurological diagnoses and exposure to the chemicals in the water at Camp Lejeune.
B4—Applicant has a psychiatric diagnosis that causes neurobehavioral symptoms. The 2003 IOM review of solvent exposures and the 2009 NRC review found inadequate or insufficient evidence of an association between these psychiatric diagnoses and exposure to the chemicals in the water at Camp Lejeune.
B5—Applicant has evidence of neurobehavioral symptoms whose onset occurred during or around the applicant's exposure at Camp Lejeune. Chronic, intermittent, or persistent symptoms since exposure suggests neurobehavioral effects secondary to exposure at Camp Lejeune. Applicant accepted into the Camp Lejeune program.

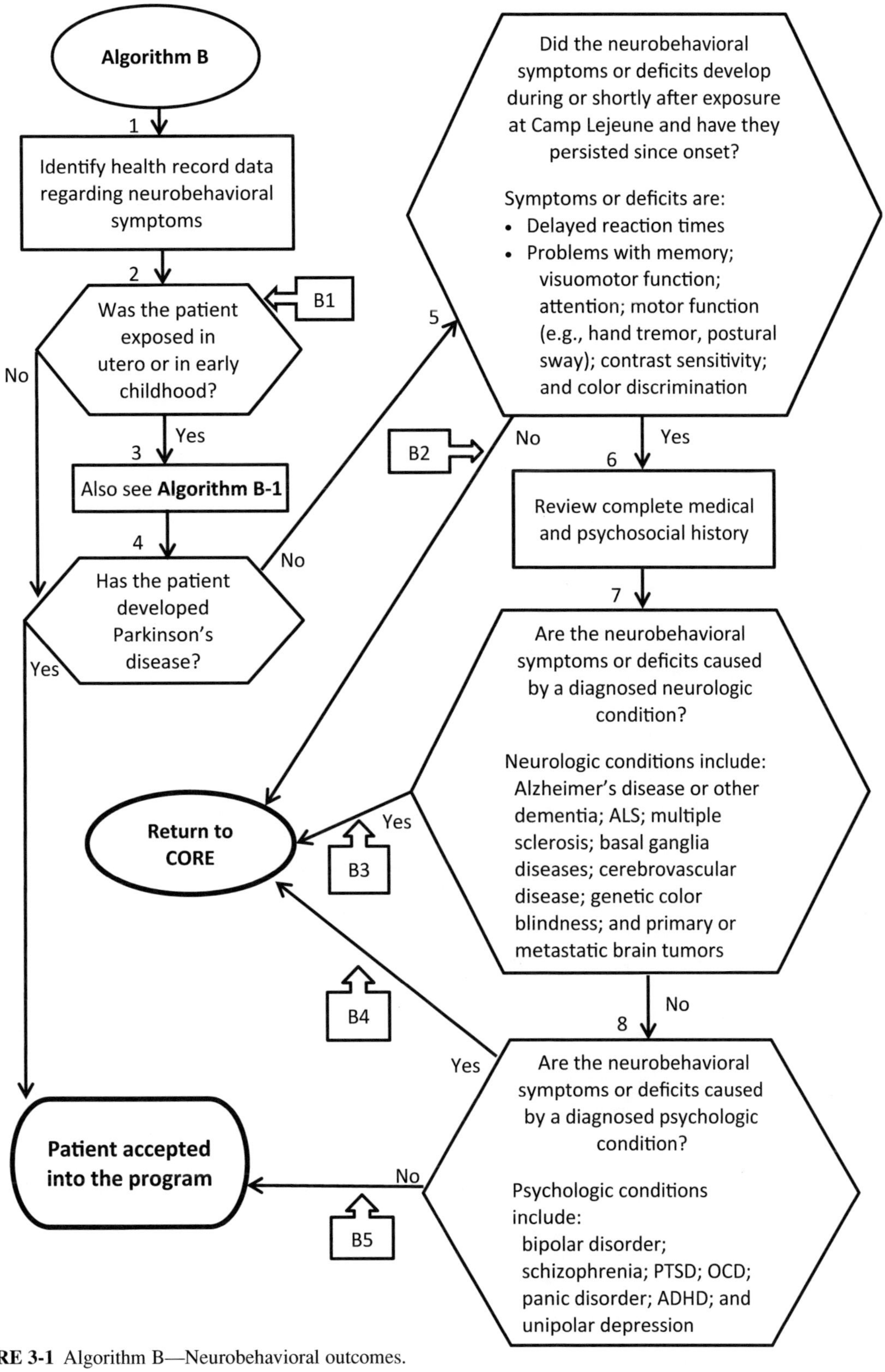

FIGURE 3-1 Algorithm B—Neurobehavioral outcomes.

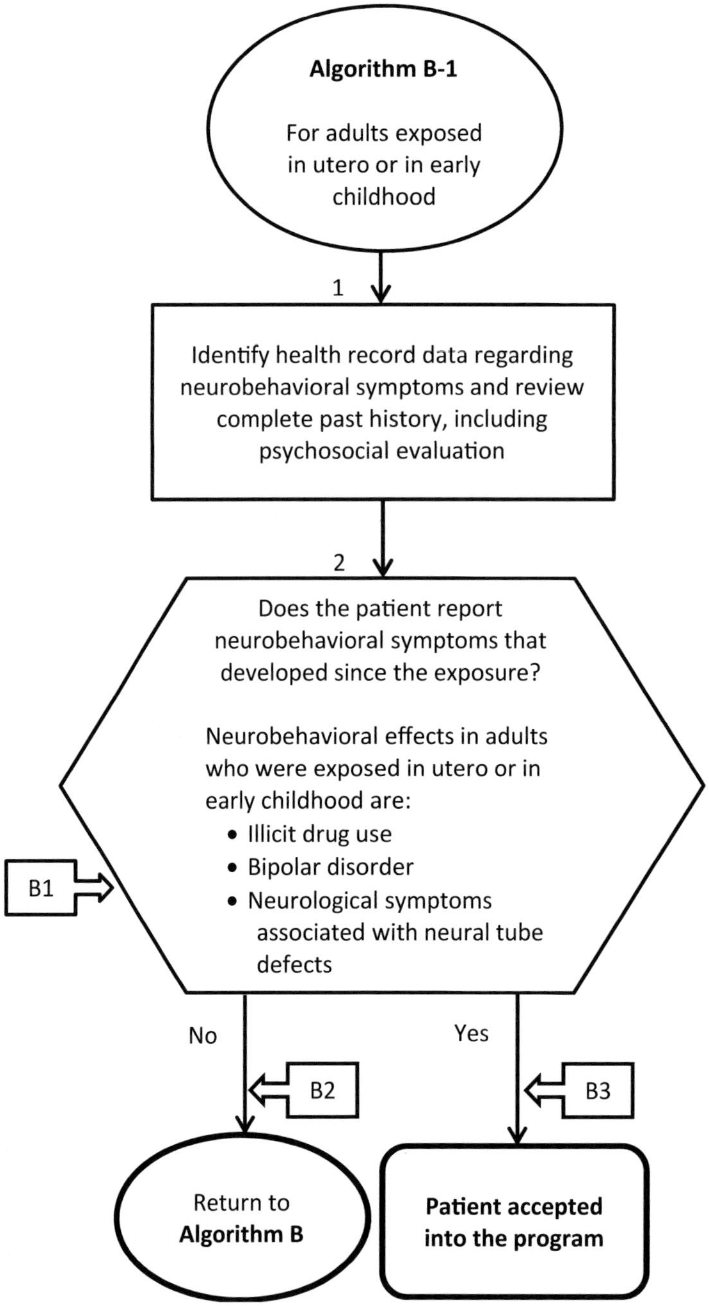

FIGURE 3-2 Algorithm B-1—Neurobehavioral outcomes in adults exposed in utero or in early childhood.

ANNOTATIONS FOR ALGORITHM B-1:

B1—Identified neurobehavioral symptoms include illicit drug use, bipolar disorder, and neurological problems associated with neural tube defects, although not all committee members were in agreement on the inclusion of illicit drug use and bipolar disorder.

B2—Applicant does not have neurobehavioral symptoms as a covered condition and is not accepted to the Camp Lejeune program at this time.

B3—Applicant accepted into the Camp Lejeune program.

REFERENCES

Aschengrau, A., J. M. Weinberg, P. A. Janulewicz, L. G. Gallagher, M. R. Winter, V. M. Vieira, T. F. Webster, and D. M. Ozonoff. 2009. Prenatal exposure to tetrachloroethylene-contaminated drinking water and the risk of congenital anomalies: A retrospective cohort study. *Environmental Health* 8:44.

Aschengrau, A., J. M. Weinberg, P. A. Janulewicz, M. E. Romano, L. G. Gallagher, M. R. Winter, B. R. Martin, V. M. Vieira, T. F. Webster, R. F. White, and D. M. Ozonoff. 2011. Affinity for risky behaviors following prenatal and early childhood exposure to tetrachloroethylene (PCE)-contaminated drinking water: A retrospective cohort study. *Environmental Health* 10:102.

Aschengrau, A., J. M. Weinberg, P. A. Janulewicz, M. E. Romano, L. G. Gallagher, M. R. Winter, B. R. Martin, V. M. Vieira, T. F. Webster, R. F. White, and D. M. Ozonoff. 2012. Occurrence of mental illness following prenatal and early childhood exposure to tetrachloroethylene (PCE)-contaminated drinking water: A retrospective cohort study. *Environmental Health* 11:2.

ATSDR (Agency for Toxic Substances and Disease Registry). 1997. *Toxicological profile for tetrachloroethylene.* U.S. Department of Health and Human Services, Public Health Service, Agency for Toxic Substances and Disease Registry.

Bale, A. S., S. Barone, Jr., C. S. Scott, and G. S. Cooper. 2011. A review of potential neurotoxic mechanisms among three chlorinated organic solvents. *Toxicology and Applied Pharmacology* 255(1):113-126.

Bove, F. J., P. Z. Ruckart, M. Maslia, and T. C. Larson. 2014a. Evaluation of mortality among marines and Navy personnel exposed to contaminated drinking water at USMC Base Camp Lejeune: A retrospective cohort study. *Environmental Health* 13(1):10.

Bove, F. J., P. Z. Ruckart, M. Maslia, and T. C. Larson. 2014b. Mortality study of civilian employees exposed to contaminated drinking water at USMC Base Camp Lejeune: A retrospective cohort study. *Environmental Health* 13:68.

Brain Development Cooperative Group. 2012. Total and regional brain volumes in a population-based normative sample from 4 to 18 years: The NIH MRI Study of Normal Brain Development. *Cerebral Cortex* 22(1):1-12.

Chen, R., F. Dick, S. Semple, A. Seaton, and L. G. Walker. 2001. Exposure to organic solvents and personality. *Occupational and Environmental Medicine* 58(1):14-18.

Costa, T. L., M. T. Barboni, A. L. Moura, D. M. Bonci, M. Gualtieri, L. C. de Lima Silveira, and D. F. Ventura. 2012. Long-term occupational exposure to organic solvents affects color vision, contrast sensitivity and visual fields. *PLoS ONE* 7(8):e42961.

Desrosiers, T. A., C. C. Lawson, R. E. Meyer, D. B. Richardson, J. L. Daniels, M. A. Waters, E. van Wijngaarden, P. H. Langlois, P. A. Romitti, A. Correa, and A. F. Olshan. 2012. Maternal occupational exposure to organic solvents during early pregnancy and risks of neural tube defects and orofacial clefts. *Occupational and Environmental Medicine* 69(7):493-499.

Dick, F. D., G. De Palma, A. Ahmadi, N. W. Scott, G. J. Prescott, J. Bennett, S. Semple, S. Dick, C. Counsell, P. Mozzoni, N. Haites, S. B. Wettinger, A. Mutti, M. Otelea, A. Seaton, P. Soderkvist, A. Felice. 2007. Environmental risk factors for Parkinson's disease and parkinsonism: The Geoparkinson study. *Occupational and Environmental Medicine* 64(10):666-672.

Dick, F. D., V. J. Bourne, S. E. Semple, H. C. Fox, B. G. Miller, I. J. Deary, and L. J. Whalley. 2010. Solvent exposure and cognitive ability at age 67: A follow-up study of the 1947 Scottish Mental Survey. *Occupational and Environmental Medicine* 67(6):401-407.

EPA (U.S. Environmental Protection Agency). 2004. *Summary report of the peer review workshop on the neurotoxicity of tetrachloroethylene.* EPA/600/R-04/041. Washington, DC: National Center for Environmental Assessment, Office of Research and Development.

EPA. 2011. *Toxicological review of tricholorethylene (CAS No. 79-01-6) in support of summary information on the Integrated Risk Information System (IRIS).* EPA/635/R-08/011F. U.S. Environmental Protection Agency.

EPA. 2012. *Toxicological review of tetracholorethylene (Perchloroethylene) (CAS No. 127-18-4) in support of summary information on the Integrated Risk Information System (IRIS).* EPA/635/R-08/011F. U.S. Environmental Protection Agency.

Flodin, U., C. Edling, and O. Axelson. 1984. Clinical-studies of psychoorganic syndromes among workers with exposure to solvents. *American Journal of Industrial Medicine* 5(4):287-295.

Gash, D. M., K. Rutland, N. L. Hudson, P. G. Sullivan, G. Bing, W. A. Cass, J. D. Pandya, M. Liu, D. Y. Choi, R. L. Hunter, G. A. Gerhardt, C. D. Smith, J. T. Slevin, and T. S. Prince. 2008. Trichloroethylene: Parkinsonism and complex 1 mitochondrial neurotoxicity. *Annals of Neurology* 63(2):184-192.

Getz, K. D., P. A. Janulewicz, S. Rowe, J. M. Weinberg, M. R. Winter, B. R. Martin, V. M. Vieira, R. F. White, and A. Aschengrau. 2012. Prenatal and early childhood exposure to tetrachloroethylene and adult vision. *Environmental Health Perspectives* 120(9):1327-1332.

Gobba, F., and A. Cavalleri. 2000. Evolution of color vision loss induced by occupational exposure to chemicals. *Neurotoxicology* 21(5):777-781.

Goldman, S. M., P. J. Quinlan, G. W. Ross, C. Marras, C. Meng, G. S. Bhudhikanok, K. Comyns, M. Korell, A. R. Chade, M. Kasten, B. Priestley, K. L. Chou, H. H. Fernandez, F. Cambi, J. W. Langston, and C. M. Tanner. 2012. Solvent exposures and Parkinson's disease risk in twins. *Annals of Neurology* 71(6):776-784.

IOM (Institute of Medicine). 1996. *Veterans and Agent Orange: Update 1996*. Washington, DC: National Academy Press.

IOM. 2003. *Gulf War and health, Volume 2: Insecticides and solvents*. Washington, DC: The National Academies Press.

Janulewicz, P. A., R. F. White, M. R. Winter, J. M. Weinberg, L. E. Gallagher, V. Vieira, T. F. Webster, and A. Aschengrau. 2008. Risk of learning and behavioral disorders following prenatal and early postnatal exposure to tetrachloroethylene (PCE)-contaminated drinking water. *Neurotoxicology and Teratology* 30(3):175-185.

Janulewicz, P. A., R. F. White, B. M. Martin, M. R. Winter, J. M. Weinberg, V. Vieira, and A. Aschengrau. 2012. Adult neuropsychological performance following prenatal and early postnatal exposure to tetrachloroethylene (PCE)-contaminated drinking water. *Neurotoxicology and Teratology* 34(3):350-359.

Janulewicz, P. A., R. J. Killiany, R. F. White, B. M. Martin, M. R. Winter, J. M. Weinberg, and A. Aschengrau. 2013. Structural magnetic resonance imaging in an adult cohort following prenatal and early postnatal exposure to tetrachloroethylene (PCE)-contaminated drinking water. *Neurotoxicology and Teratology* 38:13-20.

Lock, E. A., J. Zhang, and H. Checkoway. 2013. Solvents and Parkinson disease: A systematic review of toxicological and epidemiological evidence. *Toxicology and Applied Pharmacology* 266(3):345-355.

McDonnell, L., C. Maginnis, S. Lewis, N. Pickering, M. Antoniak, R. Hubbard, I. Lawson, and J. Britton. 2003. Occupational exposure to solvents and metals and Parkinson's disease. *Neurology* 61(5):716-717.

Murata, K., O. Inoue, M. Akutsu, and T. Iwata. 2010. Neuromotor effects of short-term and long-term exposures to trichloroethylene in workers. *American Journal of Industrial Medicine* 53(9):915-921.

National Parkinson Foundation. 2014. PD 101 for patients. Available at http://www.parkinson.org/Patients/PD-101.aspx (accessed September 11, 2014).

NIOSH (National Institute for Occupational Safety and Health). 1987. *Organic Solvent Neurotoxicity. Current Intelligence Bulletin 48*. DHHS Publication Number 87-104. Available at www.cdc.gov/niosh/docs/87-104 (accessed December 31, 2014).

NRC (National Research Council). 2009. *Contaminated water supplies at Camp Lejeune: Assessing potential health effects*. Washington, DC: The National Academies Press.

Reif, J. S., J. B. Burch, J. R. Nuckols, L. Metzger, D. Ellington, and W. K. Anger. 2003. Neurobehavioral effects of exposure to trichloroethylene through a municipal water supply. *Environmental Research* 93(3):248-258.

Ruckart, P. Z., F. J. Bove, and M. Maslia. 2013. Evaluation of exposure to contaminated drinking water and specific birth defects and childhood cancers at Marine Corps Base Camp Lejeune, North Carolina: A case-control study. *Environmental Health: A Global Access Science Source* 12(1):104.

Sanchez, C. E., J. E. Richards, and C. R. Almli. 2012. Neurodevelopmental MRI brain templates for children from 2 weeks to 4 years of age. *Developmental Psychobiology* 54(1):77-91.

Storm, J. E., K. A. Mazor, K. M. Aldous, B. C. Blount, S. E. Brodie, and J. B. Serle. 2011. Visual contrast sensitivity in children exposed to tetrachloroethylene. *Archives of Environmental and Occupational Health* 66(3):166-177.

Till, C., C. A. Westall, G. Koren, I. Nulman, and J. F. Rovet. 2005. Vision abnormalities in young children exposed prenatally to organic solvents. *Neurotoxicology* 26(4):599-613.

Van Valen, E., E. Wekking, G. van der Laan, M. Sprangers, and F. van Dijk. 2009. The course of chronic solvent induced encephalopathy: A systematic review. *Neurotoxicology* 30(6):1172-1186.

4

Other Health Outcomes

In addition to the renal toxicity and neurobehavioral effects discussed in the preceding chapters, the U.S. Department of Veterans Affairs (VA) will pay for the treatment of 13 other medical conditions specified in the Janey Ensminger Act. These conditions are detailed in the VA guidance and accompanying algorithms to help clinicians and administrators make decisions about whether or not veterans and family members are eligible for health care benefits under the Camp Lejeune Program. These other outcomes are presented here in the same order as in the guidance: cancer, scleroderma, miscarriage and infertility, and hepatic steatosis. The description and discussion of each outcome includes a brief overview, a review of the documentation and algorithm, and the committee's recommendations for improvement. Where applicable, algorithms have been revised to highlight the committee's suggested changes. Discussion of VA's decision-making process for the health conditions listed in the act, including screening and secondary health conditions, is presented in Chapter 5.

CANCER AND RELATED CONDITIONS

Overview

The Janey Ensminger Act lists eight malignant neoplasms (esophageal cancer, lung cancer, breast cancer, bladder cancer, kidney cancer, leukemia, multiple myeloma, and non-Hodgkin's lymphoma) and myelodysplastic syndromes among the 15 conditions covered by the act. These conditions were included in the legislation because of evidence presented in the 2009 National Research Council (NRC) report, which concluded that there was "limited/suggestive evidence of an association" between chronic exposure to solvents, particularly perchloroethylene (PCE), and cancers of the breast, bladder, kidneys, esophagus, and lungs (NRC, 2009). The toxicologic evidence was strongest for the associations between trichloroethylene (TCE) and kidney cancer and between PCE and kidney cancer. The report found that there was limited/suggestive evidence of an association between solvent mixtures and adult leukemia, myelodysplastic syndromes, and multiple myeloma. The report also concluded that there was inadequate/insufficient evidence to determine whether an association exists between non-Hodgkin's lymphoma and TCE, PCE, or solvent exposure.

New Research

Reports and meta-analyses published since the 2009 NRC report have also assessed whether exposure to TCE or PCE results in an increased risk of cancer, including dying from a cancer (Christensen et al., 2013; Hansen et al., 2013; Lipworth et al., 2011; Scott and Jinot, 2011). These new studies have generally supported the conclusions of the previous report and also addressed other cancers.

Because a number of authoritative reviews of the carcinogenicity of various solvents have been published since the NRC report in 2009, the committee summarizes the conclusions of those reviews in the following sections. These documents were written or reviewed by panels of experts, and the information in them was collected, analyzed, and presented systematically. The committee also reviewed two studies conducted by Agency for Toxic Substances and Disease Registry (ATSDR) of military and civilian cohorts exposed to contaminated water at Camp Lejeune (Bove et al., 2014a,b).

Authoritative Reviews

The U.S. Environmental Protection Agency (EPA) published toxicologic reviews of TCE and PCE in 2011 and 2012, respectively. Based on human studies, EPA found that TCE is carcinogenic by all routes of exposure (ingestion, inhalation, etc.) with clear evidence of a causal relationship between TCE and kidney cancer in humans. Evidence was strong for an association between TCE and non-Hodgkin's lymphoma but less so for liver and biliary cancer, esophageal, prostate, cervical, breast, and childhood cancers (EPA, 2011). EPA also concluded that PCE is likely to cause cancer in humans by all routes of exposure. This conclusion is supported by suggestive evidence in humans and by conclusive evidence in animals. Epidemiologic studies show associations between PCE and bladder cancer, non-Hodgkin's lymphoma, and multiple myeloma. More limited data exist for esophageal, kidney, lung, cervical, and breast cancers (EPA, 2012).

In 2013, the International Agency for Research on Cancer (IARC) published assessments of the association of TCE and PCE with cancer. Human studies were used to determine what specific kinds of cancer each solvent caused. Regarding TCE, IARC found sufficient evidence in humans and animals to conclude that it causes cancer. IARC found that there is sufficient evidence that TCE causes kidney cancer and identified a positive association between TCE and non-Hodgkin's lymphoma and liver cancer. It also reported statistically significant excess risks of lung, cervix, and esophageal cancers, but the evidence was insufficient to allow for specific associations to be made. Regarding PCE, IARC concluded from sufficient evidence in animals and limited evidence in humans that it is probably carcinogenic in humans. Human data show a positive association between PCE and bladder cancer, but the evidence was inconsistent for non-Hodgkin's lymphoma, and esophageal, kidney, and cervical cancers (IARC, 2014a,b).

ATSDR Studies

ATSDR has published three epidemiologic studies investigating cancers in the Camp Lejeune population exposed to contaminated drinking water. Ruckart et al. (2013) looked at increased risks of cancer among 12,598 children born to mothers residing at Camp Lejeune. The study reported eleven cases of leukemia and two cases of non-Hodgkin's lymphoma in children exposed while in utero. The odds ratios (ORs) for any childhood cancer were elevated but not statistically significant for first-trimester exposure to PCE, vinyl chloride, and dichloroethylene. This study was limited by the small numbers of events and the possibility of differential recall bias and missing data. Bove et al. (2014a) compared the mortality of Marine Corps and Navy personnel exposed to contaminated drinking water at Camp Lejeune with that of Marine Corps and Navy personnel stationed at Camp Pendleton. Elevated hazard ratios for the Camp Lejeune cohort were reported for deaths resulting from kidney cancer, liver cancer, esophageal cancer, cervical cancer, multiple myeloma, and Hodgkin lymphoma; however, the confidence intervals all included 1.0 (Bove et al., 2014a). The same authors conducted an almost identical analysis among civilian workers employed at Camp Lejeune and Camp Pendleton. Elevated but statistically nonsignificant hazard ratios for the Camp Lejeune cohort for kidney cancer, hematopoietic cancers, multiple myeloma, leukemias, rectal cancer, lung cancer, and oral cancer were observed among the 197 total deaths at Camp Lejeune and 234 at Camp Pendleton (Bove et al., 2014b). The authors concluded that a longer follow up would be necessary in both cohorts to allow for more precise estimates because only 6% of the military cohort and 14% of the civilian cohort had died at the time of the assessment (Bove et al., 2014a,b).

Latency

Adult cancers have latency periods of years and childhood cancers have latency periods of at least months (de Gonzalez et al., 2012). The Centers for Disease Control and Prevention's World Trade Center Health Program has summarized the latency periods for various cancers, suggesting a minimum latency of 4 to 20 years for most solid tumors and several months to 10 or 15 years for lymphoproliferative and hematopoietic cancers (Howard, 2013). The VA guidance and algorithm do not explicitly consider latency in coverage determinations for the cancer diagnoses.

VA Guidance and Core Algorithm

For the eight cancers and myelodysplastic syndromes covered by the act, there are well-established diagnostic criteria. The Camp Lejeune program includes additional health care benefits in the form of comprehensive medical care during active cancer treatment, because such treatment (including surgery, chemotherapy, and radiation) can result in systemic secondary effects on virtually all other organ systems. The treating oncologist may specify the active treatment time period or VA will provide coverage in 6-month increments after the initial diagnosis.

The core algorithm in the guidance is used for all the cancers regardless of other risk factors and time of onset. (Onset, latency periods, and exposures for cancer are only briefly mentioned in the guidance on page 7.) Because the diagnosis of these cancers is expected to be based on established criteria and because the risk attributable to other causes cannot be ascertained, all qualified veterans and family members with one of these diagnoses are automatically accepted to the Camp Lejeune program and are eligible for coverage (Walters, 2014a).

Recommendations

The core clinical algorithm addresses cancer diagnoses, asking whether the veteran or family member has an established diagnosis of one of the eight cancers or myelodysplastic syndromes. Although this is relatively straightforward, VA may want to consider the following findings and recommendations:

- According to VA, it plans to cover tumors regardless of latency. This follows the precedent set by VA in response to Agent Orange exposures for Vietnam veterans, and provides the benefit of the doubt to the veteran and, in this case, the family member (Walters, 2014a).

The committee recommends that VA clearly state in the guidance its policy decision to not consider the latency of cancers.

- Second, VA may want to clarify whether it will cover second primary cancers if the first primary (which must be one of the cancers covered by the act) occurred before the exposure at Camp Lejeune.

The committee recommends that VA include in the Camp Lejeune program patients with second primary cancers (but not recurrent or metastatic cancers) whose primary cancer was one of the covered cancers, even if their first primary cancer was diagnosed before residence at Camp Lejeune.

- Third, the guidance and algorithm do not address whether precancerous lesions of the cancers covered by the act are also covered—such as ductal carcinoma in situ (noninvasive breast cancer); Barrett's esophagus, which can precede esophageal cancer; and monoclonal gammopathy of undetermined significance, which may precede multiple myeloma. VA has indicated to the committee that it plans to cover precancerous lesions (Walters, 2014b), and the committee finds this approach to be reasonable.

The committee recommends that VA clearly address precancerous lesions in the clinical guidance and in the core algorithm.

- Fourth, the guidance defines active treatment for cancer as surgery, chemotherapy, or radiation therapy, or some combination of the three, but it does not specifically include hormonal treatment or immunotherapy. Even if the primary purpose of such treatment is to prevent the recurrence of cancer (e.g., hormonal therapy to prevent recurrence of breast cancer), such treatment is indicated in 38 CFR 17.38, which states that the VA medical benefits package covers treatment to prevent recurrence of a disease.

The committee recommends that VA specifically include hormonal treatment and immunotherapy as part of the "active treatment" for cancer in the clinical guidance.

SCLERODERMA (SYSTEMIC SCLEROSIS)

Overview

Scleroderma, also referred to as systemic sclerosis, is a rare autoimmune condition characterized by the presence of thickened, sclerotic skin lesions. It is the result of an overproduction and accumulation of collagen in tissues (Scleroderma Foundation, 2014). The impact and symptoms vary from having patches of hardened skin to the involvement of other tissues and organs such as the heart, lungs, kidney, and digestive system (Scleroderma Foundation, 2014). Scleroderma is thought to be caused by several factors, with the immune system, vascular system, and connective tissue metabolism all playing a role (NORD, 2014). Scleroderma occurs most commonly in adults (Scleroderma Foundation, 2014).

Scleroderma can be broadly classified into three groups: systemic sclerosis or systemic scleroderma; localized scleroderma[1] (morphea and linear scleroderma); and scleroderma-like conditions, a heterogeneous group of diseases linked by the presence of thickened, sclerotic skin. These scleroderma-like conditions include eosinophilic fasciitis, localized forms of scleroderma, scleredema and scleromyxedema, keloids, and environmental exposure–associated conditions, including eosinophilia–myalgia syndrome and pseudosclerodermas induced by various drugs (Mori et al., 2002). CREST syndrome (calcinosis, Raynaud's phenomenon, esophageal dysfunction, sclerodactyly, and telangiectasia) is another subdisorder (ISN, 2014). Risk factors for scleroderma include abnormal immune activity, environmental triggers, and genetic predisposition (NORD, 2014).

Scleroderma has an annual incidence of 1 to 2 per 100,000 individuals in the United States (Lawrence et al., 1998). Estimates indicate that scleroderma affects between 40,000 and 165,000 people in the United States (NORD, 2014). Peak onset is between the ages of 20 and 55, and the disease is more common in women (Mayo Clinic, 2013; NORD, 2014; Scleroderma Foundation, 2014).

The diagnosis of scleroderma can be difficult, and misdiagnoses and undiagnosed cases may be common. The American College of Rheumatology has developed and supported established diagnostic criteria for scleroderma since 1980. In 2001, it published diagnostic criteria requiring that a patient have either proximal diffuse sclerosis (skin tightness, thickening, non-pitting induration) or at least two of the following three symptoms: sclerodactyly of fingers or toes, digital pitting scars or loss of substance of finger pads (pulp loss), or bilateral basilar pulmonary fibrosis. In 2013, the diagnostic criteria for systemic sclerosis were updated in collaboration with the European League Against Rheumatism. Validation of the new criteria indicated a sensitivity of 91% and specificity of 92% vs 75% and 72%, respectively, for the former criteria in the same sample of patients. The new criteria rely on

[1] Morphea occurs in adults, is characterized by having skin plaques that are oval shaped and ivory colored but with no involvement of internal organs, and generally improves without treatment. Linear scleroderma generally occurs in children and manifests as thick skin on arms or legs and can cause a limb to grow more slowly than its counterpart (NORD, 2014).

TABLE 4-1 Diagnostic Criteria for Systemic Sclerosis (Scleroderma)

Item	Subitem(s)	Weight/Score*
Skin thickening of the fingers of both hands extending proximal to the metacarpophalangeal joint *(sufficient criteria)*	—	9
Skin thickening of the fingers *(only count the higher score)*	Puffy fingers	2
	Sclerodactyly of the fingers (distal to the metacarpophalangeal joints but proximal to the proximal interphalangeal joints)	4
Fingertip lesions *(only count the higher score)*	Digital tip ulcers	2
	Fingertip pitting scars	3
Talengiectasia	—	2
Abnormal nailfold capillaries	—	2
Pulmonary arterial hypertension and/or interstitial lung disease *(maximum score of 2)*	Pulmonary arterial hypertension	2
	Interstitial lung disease	2
Raynaud's phenomenon	—	3
SSc-related autoantibodies (anti-centromere, anti-topoisomerase I [anti-Scl-70], anti-RNA polymerase III) *(maximum score is 3)*	Anti-centromere	3
	Anti-topoisomerase I	
	Anti-RNA polymerase III	

NOTE: These criteria are applicable to any patient considered for inclusion in an SSc study. The criteria are not applicable to patients with skin thickening sparing the fingers or to patients who have a scleroderma-like disorder that better explains their manifestations (e.g., nephrogenic sclerosing fibrosis, generalized morphea, eosinophilic fasciitis, scleroderma diabeticorum, scleromyxedema, erythromyalgia, porphyria, lichen sclerosis, graft-versus-host disease, diabetic cheiroarthropathy). SSc = systemic sclerosis (scleroderma).
* The total score is determined by adding the maximum weight (score) in each category. Patients with a total score of ≥ 9 are classified as having definite SSc.
SOURCE: Reproduced from van den Hoogen et al. (2013) with permission from BMJ Publishing Group Ltd.

weights for various aspects of the disease; patients with a score of 9 or greater were classified as having systemic scleroderma (van den Hoogen et al., 2013). The new criteria are shown in Table 4-1.

Other important clinical features may include dysphagia, hypertension, and renal insufficiency; diarrhea with malabsorption; dyspnea secondary to the lung involvement; mucocutaneous telangiectasia on the face, lips, oral cavity, or hands; and erectile dysfunction (Varga, 2014).

A skin biopsy is generally not essential for confirmation, but blood tests or other studies may be helpful in confirming the diagnosis, as might consultation with a dermatologist or rheumatologist (Scleroderma Foundation, 2014). Treatment is limited to symptom management and efforts to improve quality of life; there is no cure (NORD, 2014; Scleroderma Foundation, 2014).

Although scleroderma is generally considered an autoimmune disease, a variety of occupational and environmental exposures have been associated with its development, including exposure to silica, vinyl chloride, and adulterated rapeseed oil (Mora, 2009). Scleroderma has also been reported with exposure to organic solvents and epoxy resins.

The mechanisms by which solvents induce autoimmune effects are poorly understood. The specific gaps in scientific knowledge include insufficient data on human immune suppression, on the effects of age and sex on susceptibility to TCE-related autoimmune effects, and on the effects of dose, duration, and timing of exposure (Weinhold, 2009). Most of the immune alterations associated with scleroderma involve antigen recognition, cell signaling, and cytokine production, but there may be multiple mechanisms by which environmental exposures initiate or contribute to the development of scleroderma (Mora, 2009).

Epidemiologic Studies of Exposure to Organic Solvents

In 2003, the Institute of Medicine (IOM) concluded that there was inadequate/insufficient evidence to determine whether an association exists between solvent exposure and scleroderma. On the basis of four additional studies of occupational solvent exposure (IOM, 2003), in 2009 the NRC concluded that the evidence of an association between mixed solvent exposure and scleroderma is limited/suggestive with some evidence pointing toward TCE exposure in particular (NRC, 2009).

Subsequent to the 2009 NRC report, reviews conducted by authoritative entities have also noted TCE's autoimmune effects. In light of the new evidence, an expert panel of the National Institute of Environmental Health Sciences examined the epidemiologic data available at the time and concluded that solvent exposure could contribute to the development of scleroderma (Miller et al., 2012). In 2011, EPA conducted a toxicological review of TCE, which noted:

> The relation between systemic autoimmune diseases, such as scleroderma, and occupational exposure to TCE has been reported in several recent studies. A meta-analysis of scleroderma studies (Garabrant et al., 2003; Diot et al., 2002; Nietert et al., 1998) conducted by the EPA resulted in a statistically significant combined OR for any exposure in men (OR: 2.5, 95% CI: 1.1, 5.4), with a lower OR seen in women (OR: 1.2, 95% CI: 0.58, 2.6). (EPA, 2011, p. 4-427)

Two meta-analyses also examined the risk of scleroderma following solvent exposure. Barragan-Martinez et al. (2012) showed that organic solvent exposure is associated with systemic scleroderma (OR = 2.54, 95% CI [confidence interval] 1.23–5.14) and all autoimmune disorders (OR = 1.54, 95% CI 1.25–1.92) and that people with inherent risk factors (familial autoimmune disorders or genetic susceptibility) are particularly at risk. Cooper et al. (2009) used the concordance between human and animal studies to support the role of TCE in autoimmune diseases (skin hypersensitivity with systemic effects) and, based on an analysis of three case-control studies, reported an OR of 2.5 (95% CI 1.1–5.4) for scleroderma among TCE-exposed workers.

Additional literature reviews of epidemiologic evidence have supported the role of solvents or TCE as a risk factor for scleroderma (Mora, 2009) and autoimmune effects (Gilbert, 2010; Pollard, 2012; Pollard et al., 2010).

VA Guidance and Algorithm

As is the case with cancer, scleroderma's diagnostic criteria are well established. Currently the VA guidance refers to the American College of Rheumatology criteria published in 2001; however, in 2013 the American College of Rheumatology adopted new criteria for systemic sclerosis (van den Hoogen et al., 2013) (see Table 4-1).

Because onset can occur at any time after exposure to a toxicant, any exposed veteran or family member is eligible for health benefits and accepted to the Camp Lejeune program regardless of when the disease was diagnosed. The core algorithm asks if a patient has an established diagnosis of scleroderma, and if the answer is yes, the patient is accepted into the program. Again, as with cancer, there is no separate algorithm for scleroderma; it is one step in the core algorithm.

Recommendations

The committee finds that the guidance and the core algorithm for scleroderma are reasonable and appropriate.

The committee recommends that VA update the guidance in accordance with the 2013 American College Rheumatology diagnostic criteria for scleroderma.

MISCARRIAGE AND INFERTILITY

This section provides separate descriptions of miscarriage and infertility and new research on each of them, including on their long-term health consequences. The committee then reviews the guidance and algorithm on miscarriage and infertility together since VA combined these outcomes into one algorithm, algorithm W Reproductive Health: Miscarriages and Infertility in Women.

Miscarriage

Overview

Miscarriage is a common term used to describe a spontaneous abortion. A spontaneous abortion is the naturally occurring expulsion of an embryo or fetus before viability. In clinical practice, expulsion before 20 weeks gestation is defined as a miscarriage (Storck, 2012). The rate at which miscarriages occur in the general population is difficult to estimate because many go undetected and unreported. Rates also vary greatly by gestational age; in one study, the highest reported rate was more than 20 miscarriages per 1,000 women-weeks up to week 13. Rates fall steadily with each additional week of gestation. Estimates suggest that between 11% and 22% of pregnancies between weeks 5 and 20 end in miscarriage (Ammon Avalos et al., 2012).

Common causes of miscarriage include maternal hormone problems, infections, physical and emotional trauma, being of an older age (there is a 50% chance of miscarriage in women over 45), smoking, illicit drug use, malnutrition, excessive caffeine, radiation exposure, and exposure to toxic substances, including the solvents to which residents at Camp Lejeune were exposed. Women who have had previous miscarriages are also at increased risk of additional miscarriages (25%, but this risk is only slightly increased compared to women who have not previously miscarried) (American Pregnancy Association, 2014).

A miscarriage may result in long-term psychologic and medical consequences. Mental health effects reported after miscarriage include depression, anxiety, and posttraumatic stress disorder (Frost and Condon, 1996). Depression and anxiety resulting from a previous miscarriage can be persistent, do not necessarily diminish after the birth of a subsequent healthy child (Blackmore et al., 2011), and differ from other kinds of perinatal loss (Adolfsson, 2011; Broen et al., 2004). Depression after miscarriage may be accompanied by long-lasting psychological, social, and health consequences (Beutel et al., 1995).

New Research

Recent research continues to support the association between solvent exposure and miscarriage. Miscarriage has been associated with occupational exposures to solvents in many industries, including wood processing (Viragh et al., 2014), pharmaceutical production (Attarchi et al., 2012), hairdressing (Peters et al., 2010), dry cleaning, semiconductor manufacturing, and petrochemical production (Kumar, 2011). In contrast to the occupational studies, a study on a general population cohort of women exposed to PCE-contaminated drinking water in Cape Cod did not find any meaningful associations between exposure and pregnancy loss (Aschengrau et al., 2009).

EPA's toxicological review of PCE indicated that while some research is limited by imprecise estimates or an inability to evaluate confounding factors, occupational studies have generally reported maternal solvent exposure to be associated with elevated risks of miscarriage. However, studies of two populations did not observe an association (EPA, 2012). EPA's review of TCE reported the same limitations for epidemiologic studies and found them to be not "highly informative" (EPA, 2011). ATSDR's updated assessment of TCE reported little published evidence of an association between spontaneous abortion and TCE exposure (ATSDR, 2013). However, animal data show a consistent association between TCE exposure and prenatal loss (EPA, 2011).

Infertility

Overview

Infertility is defined as the failure to achieve a clinical pregnancy after 12 months or more of regular unprotected sexual intercourse. About 10% of women aged 15–44 have difficulty getting pregnant or staying pregnant (Eisenberg and Brumbaugh, 2009), but other estimates indicate a much lower prevalence (Mascarenhas et al., 2012; Thoma et al., 2013).

Female infertility may be caused by many factors affecting several different aspects of female reproduction. Polycystic ovary syndrome, hyperprolactinemia, eating disorders, excessive exercise, injury, and tumors all may affect ovulation. Uterine or cervical abnormalities, such as uterine fibroids or other tumors, may distort the uterus or block fallopian tubes. Pelvic inflammatory disease, sexually transmitted diseases, and other conditions may cause inflammation of or damage to fallopian tubes. Endometriosis may affect the fallopian tubes, uterus, and ovaries. Primary ovarian insufficiency (early menopause) can be caused by immune disease, radiation therapy or chemotherapy, and smoking. Pelvic infections or surgeries can cause pelvic adhesions (scar tissue). Various health conditions including thyroid hormone abnormalities, cancer and cancer treatments, celiac disease, Cushing's disease, sickle cell disease, kidney disease, diabetes, and genetic abnormalities, as well as certain medications can also reduce a woman's fertility. Reduced fertility in women is also related to age, being overweight or underweight, and the use of tobacco and alcohol (Mayo Clinic, 2014). Exposure to environmental contaminants—and, in particular, PCE and TCE—can also affect fertility, for example by reducing fecundity and altering menstrual cycles (Dzubow et al., 2010; EPA, 2011).

Long-Term Effects of Infertility

Infertility may have an impact on the quality of life and mental health. Some studies indicate that infertility is associated with depression and loneliness, as well as with social isolation in older women (although the association may be confounded by marital status) (Gift and Spence, 2014). Psychological distress caused by infertility may be exacerbated by an extended duration and by infertility treatments (Greil, 1997).

VA Guidance and Algorithm

The guidance for female infertility and miscarriage is more specific with regard to the onset and occurrence of these effects than it is for other outcomes. Exposed veterans and family members who experienced or were diagnosed with these problems during their time at Camp Lejeune are eligible for health benefits if they require ongoing medical treatment. The guidance clearly states that there is no evidence to support an increased risk of female infertility or miscarriage after the exposure ended or after an individual moved away from Camp Lejeune. Thus, current infertility or a miscarriage in a woman who was a child, adolescent, or young adult while at Camp Lejeune is excluded.

Furthermore, the guidance states that there is no evidence that the exposure of a fetus to solvents increases the risk of that person being infertile when he or she becomes a reproductively mature adult. This excludes any claims of miscarriage or infertility from the offspring of women who were pregnant while exposed to contaminated drinking water at Camp Lejeune between 1957 and 1987.

The guidance focuses on women with complications secondary to past infertility or miscarriage that require continued treatment. These complications would have developed or persisted between 26 and 56 years after problems with infertility and miscarriage occurred while at Camp Lejeune. Algorithm W begins with the identification of past health record data for pregnancy, miscarriage, or infertility dating to 1957–1987. There must be documentation that the infertility or miscarriage occurred during residence on Camp Lejeune. It is possible that medical records from that period may not be available. In these cases, it is important that VA encourage informed clinical judgment to identify veterans or family members with persistent problems that may have resulted from miscarriage or infertility that occurred concurrent with exposure to drinking water at Camp Lejeune. The next step in algorithm W asks the clinician to determine if there are ongoing "medical complications" or "medical problems" that might be attributed to the infertility or miscarriage that occurred at that time. Those health conditions, including mental health problems, are covered by the Camp Lejeune program if they can be related to miscarriage or infertility that occurred while in residence at Camp Lejeune.

Recommendations

The committee finds the guidance and algorithm for miscarriage and infertility to be generally appropriate. Suggested revisions to algorithm W are shown in Figure 4-1.

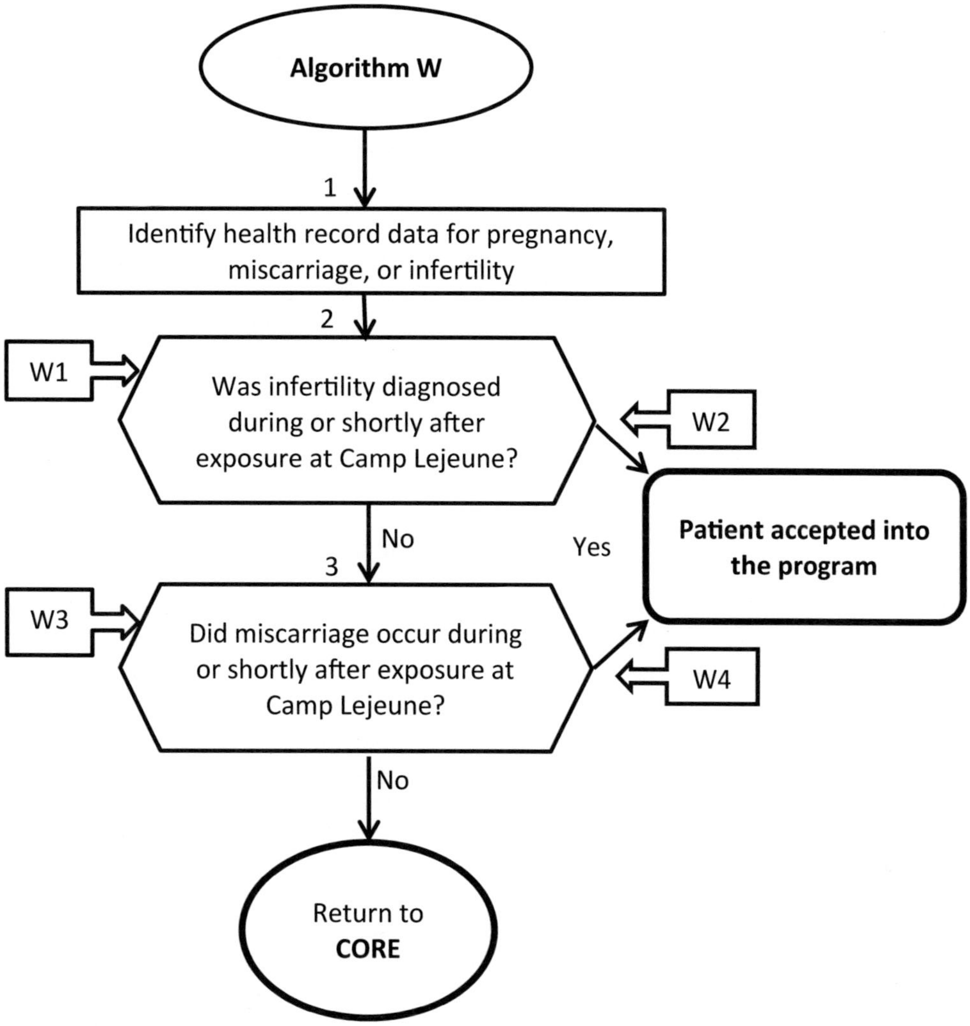

FIGURE 4-1 Revised algorithm W—Reproductive health: miscarriage and infertility in women.
ANNOTATIONS FOR ALGORITHM W:

W1—Infertility was diagnosed after leaving Camp Lejeune. There is currently no scientific evidence to support an association with chronic female infertility after cessation of exposure to solvents. Similarly the NRC report found no evidence that exposure to organic solvents while in utero increases the risk for adverse fertility effects as a reproductively mature adult. Applicant does not have female infertility that is covered by the Camp Lejeune program.

W2—Applicant has a physical or mental health condition requiring continued medical treatment from female infertility that occurred while being exposed to contaminated water at Camp Lejeune. The medical condition is related to the infertility experienced during residence at Camp Lejeune. Applicant accepted into the Camp Lejeune program.

W3—Miscarriage occurred after leaving Camp Lejeune. Current scientific evidence suggests that there are no persistent effects of solvent exposure on miscarriage or fetal loss. Applicant does not have a miscarriage that is covered by the Camp Lejeune program. Applicant is not accepted into the Camp Lejeune program at this time.

W4—Applicant has a physical or mental health condition requiring continued medical treatment from a miscarriage experienced while exposed to contaminated water at Camp Lejeune. Clinicians should carefully assess whether continued health care is needed for chronic, persistent medical problems associated with a miscarriage that occurred during solvent exposure at Camp Lejeune; if care needs are persistent, the applicant is accepted into the Camp Lejeune program.

The committee recommends that throughout the guidance and algorithm VA refer to "physical and mental health conditions" related to prior infertility or miscarriage, rather than "medical conditions," "medical problem," or "medical treatment."

HEPATIC STEATOSIS

Overview

Hepatic steatosis, commonly referred to as fatty liver, is the initial pathologic manifestation of fatty liver disease. It is an accumulation of lipids in hepatocytes (Day, 2006; Kaiser et al., 2012), which leads to an inflammatory response in the liver which in turn may progress to fibrosis, cirrhosis (Wahlang et al., 2013), and liver cancer (Du and Wang, 1998; Jiang et al., 2014). Hepatic steatosis is associated with a variety of other conditions, including type 2 diabetes; metabolic syndrome; hepatitis; hyperlipidemia; other less common liver diseases, such as Weber–Christian syndrome, Wilson disease, and lipodystrophy (Angulo, 2002; Bayard et al., 2006); starvation (McAvoy et al., 2006); severe weight loss such as experienced after bariatric surgery (Mohanty, 2006); and increased cardiovascular risk (Anstee et al., 2013; Bhatia et al., 2012a,b). Hepatic steatosis may also result from the use of some medications (e.g., chemotherapeutic agents) and from exposure to some solvents (such as TCE, PCE, and chloroform), halogenated hydrocarbons (carbon tetrachloride, vinyl chloride), volatile organic mixtures, pesticides, and nitro-organic compounds (Wahlang et al., 2013). Susceptibility to hepatic steatosis is influenced by a number of factors including genetics, alcohol consumption, the use of prescription medications, and nutritional factors such as obesity (Wahlang et al., 2013). Steatosis occurs in 90% of those who consume 16 g of alcohol or more per day (Wahlang et al., 2013). (Note: One "standard" drink [e.g., 5 ounces of wine, 12 ounces of regular beer, 7–8 ounces of malt beer, or 1.5 ounces of 80-proof spirits] contains roughly 14 grams of pure alcohol.[2]) Hepatic steatosis may also occur in up to 92% of obese adults (WGO, 2012).

Drugs reported to cause fatty liver include methotrexate, tamoxifen, corticosteroids, griseofulvin, diltiazem, anti-retroviral therapy, nifedipine (Mohanty, 2006), valproate (Depakote), high doses of intravenous tetracycline or amiodarone, and certain herbs (for example, the Chinese herb jin bu huan, used as a sedative and pain reliever) (Lee, 2014).

Most patients who present with hepatic steatosis have elevated liver enzyme levels but may be asymptomatic; others may complain of fatigue and right upper quadrant abdominal fullness or pain. Up to 50% of patients with steatosis have hepatomegaly (Sanyal, 2002). A probable diagnosis can be established by imaging studies including ultrasound, computerized tomography (CT), and magnetic resonance imaging (MRI). A definitive diagnosis is established by a pathological finding on liver biopsy, accompanied by the exclusion of other potential causes (Angulo, 2002).

Treatment focuses on removing the inciting agent (such as alcohol or a toxic exposure) or reducing risk factors related to associated conditions, including weight loss, exercise, and the treatment of diabetes and hyperlipidemia. No medications intended to protect hepatocytes have been found to be effective in treating steatosis (e.g., ursodexycholic acid, vitamin E, betaine) (Bayard et al., 2006) and there are no U.S. Food and Drug Administration–approved medications for the treatment of fatty liver disease. The prognosis for hepatic steatosis is generally benign and the condition reversible, but if it persists, more severe pathologies can develop, such as fibrosis, cirrhosis (such as cryptogenic cirrhosis), and liver cancer (Kaiser et al., 2012; McAvoy et al., 2006).

VA Guidance and Algorithm

The VA guidance states that in evaluating whether a veteran or a family member has hepatic steatosis that may be the result of exposure to drinking water at Camp Lejeune, the clinician should first consider whether it is more likely than not that the patient's hepatic steatosis is the result of a "known etiology." The most common causes include obesity and significant alcohol consumption; less common causes include dyslipidemia; metabolic

[2] See http://niaaa.nih.gov/alcohol-health/overview-alcohol-consumption/standard-drink (accessed November 5, 2014).

syndrome; diabetes; hepatitis; other liver diseases such as hepatitis C genotype 3 and Wilson's disease; and some medications.

Algorithm H begins with a diagnosis of hepatic steatosis based on medical records and liver biopsy or evidence from imaging studies (e.g., ultrasound, CT, or MRI). The committee notes that ideally a clinician would also evaluate any abnormal liver function results from tests that were conducted while the individual resided at Camp Lejeune or shortly after; however, the committee emphasizes that these test results are unlikely to be available because toxicant-induced injury soon after exposure would likely have been asymptomatic, and such tests would not have been conducted.

The guidance states that hepatic steatosis may occur during or shortly after acute exposure to solvents. In most cases hepatic steatosis associated with solvent exposure is expected to resolve after the exposure ceases; and its onset is unlikely to occur many months or years later. Thus, the guidance suggests that clinicians carefully consider the onset and duration of the condition when evaluating its potential association with drinking water at Camp Lejeune. However, neither the guidance nor algorithm H provide further information on how to assess whether the onset and duration support an association between steatosis and exposure to solvents at Camp Lejeune. The committee notes that the published literature does not provide definitive evidence regarding the onset and duration of steatosis and its resolution after the exposure ends.

The guidance specifies that if a patient's history is consistent with a known cause of hepatic steatosis, the patient would not be covered by the Camp Lejeune program. Conversely, Camp Lejeune veterans and family members with hepatic steatosis of unclear or unknown etiology should be covered by the program. The guidance on hepatic steatosis concludes by noting that "if a patient's clinical course is atypical or progresses faster than expected, then exacerbation by TCE, PCE or other organic solvents from Camp Lejeune should be considered" (page 10). The committee was unable to find evidence to support this statement.

The third step in algorithm H asks the clinician to determine if there are diagnoses or medical record data indicating other likely etiologies for the hepatic steatosis. Possible causes to be considered include obesity, alcohol abuse, dyslipidemia, metabolic syndrome, diabetes, hepatitis, and other liver diseases such as hepatitis C genotype 3 and Wilson's disease. Medical record data to be considered by the clinician include elevated body mass index, a history of hepatitis, alcohol use, and liver disease or metabolic syndrome. If a patient's hepatic steatosis is not consistent with those other possible causes, then the patient is accepted to the Camp Lejeune program.

In considering other etiologies, the algorithm suggests a threshold of 20 grams of alcohol per day in women and 30 grams per day in men to indicate alcohol-related fatty liver disease (AFLD). Unfortunately, AFLD is pathologically indistinguishable from non-alcoholic fatty liver disease, or toxicant-associated fatty liver disease (TAFLD) (Wahlang et al., 2013). To distinguish AFLD from other types of fatty liver disease, two cutoffs based on alcohol consumption have been proposed: two drinks per day or greater (Bayard et al., 2006), or 20 g alcohol/day for women and 30 g alcohol/day for men (Adams et al., 2005). Neither TAFLD nor toxicant-associated steatohepatitis—a more severe form of TAFLD characterized by hepatic steatosis, inflammatory infiltrate, and in some cases fibrosis—is associated with significant alcohol consumption or obesity.

Recommendations

The application of algorithm H for hepatic steatosis is challenging because of the high prevalence of other potential causes of hepatic steatosis in the general population, such as obesity, alcohol use, diabetes, dyslipidemia, medications, and other exposures. Thus, it is important that VA encourage informed clinical judgment to identify veterans or family members with hepatic steatosis that may have resulted from exposure to drinking water at Camp Lejeune based on its persistence since residing at Camp Lejeune or the absence of other more likely causes. In contrast, when other causes are present, steatosis is not likely to be attributable to exposure at Camp Lejeune.

The phrase in the guidance "[M]oreover if a patient's clinical course is atypical or progresses faster than expected, then exacerbation by TCE, PCE or other organic solvents from Camp Lejeune should be considered" is not consistent with the known pathogenesis of TAFLD. In particular, there is no evidence that solvent exposure would result in an atypical presentation or rapid progression of hepatic steatosis at a later date.

Based on the evidence, the committee recommends that VA delete the phrase "atypical or progresses faster than expected" in the clinical guidance. The committee further recommends that VA replace the term "alcohol abuse," listed among the other causes of hepatic steatosis in the clinical guidance and algorithm, with "alcohol use ≥ 20 g/day for women or ≥ 30 g/d for men."

There are several commonly used medications that are known to cause fatty liver, including methotrexate, tamoxifen, corticosteroids, griseofulvin, diltiazem, anti-retroviral therapy, amiodarone, nifedipine, and valproate.

The committee recommends that VA include "some medications" in the list of other causes in algorithm H and that examples of those medications be listed in the text of the clinical guidance.

Suggested revisions are shown in Figure 4-2.

ANNOTATIONS FOR ALGORITHM H:
H1—Hepatic steatosis (fatty liver) is an accumulation of lipids (triglycerides and other lipids) in the liver hepatocytes. Patients are often asymptomatic. It is often diagnosed as an incidental finding on routine medical exams with blood tests revealing abnormal liver function tests. Liver biopsy is the only definitive test to confirm diagnosis, exclude other causes, assess extent and predict prognosis. In most instances, it will be possible to identify the existence of hepatic steatosis and to define the extent of the condition using diagnostic imaging techniques (ultrasound, CT, or MRI).
H2—Applicant does not have clinical evidence (positive biopsy, CT, MRI, or ultrasound test) of hepatic steatosis at this time.
H3—The most common known causes of steatosis are obesity and alcohol abuse. Other possible causes include dyslipidemia, metabolic syndrome, diabetes, hepatitis or other liver disease, and some medications such as methotrexate, tamoxifen, corticosteroids, griseofulvin, diltiazem, anti-retroviral therapy, amiodarone, nifedipine, and valproate.
Applicant has clinical evidence of hepatic steatosis due to a cause other than exposure at Camp Lejeune. Fatty liver disease can be divided into two main categories: alcohol-related fatty liver disease and non-alcoholic fatty liver disease (NAFLD). Consumption of < 20 gm alcohol per day in women and < 30 gm in men suggests a diagnosis of NAFLD. NAFLD is associated with obesity and with abnormal glucose tolerance and dyslipidemia, and has been described as the hepatic manifestation of the metabolic syndrome. Fatty liver develops in 46% to 90% of heavy alcohol users, and in up to 94% of obese individuals. Thus, hepatic steatosis in women who drink ≥ 20 g of alcohol or in men who drink ≥ 30 g of alcohol per day or who are obese (that is, have a body mass index of > 30), is not likely due to exposure to the contaminated water at Camp Lejeune. A typical drink contains 14 g of alcohol (http://www.niaaa.nih.gov/alcohol-health/overview-alcohol-consumption/standard-drink [accessed November 5, 2014]).
Current hepatic steatosis is due to another cause other than exposure to contaminated water at Camp Lejeune. Applicant does not have a condition eligible for coverage by the Camp Lejeune program at this time.
H4—Applicant has hepatic steatosis of unknown etiology. Applicant is accepted into the Camp Lejeune program.

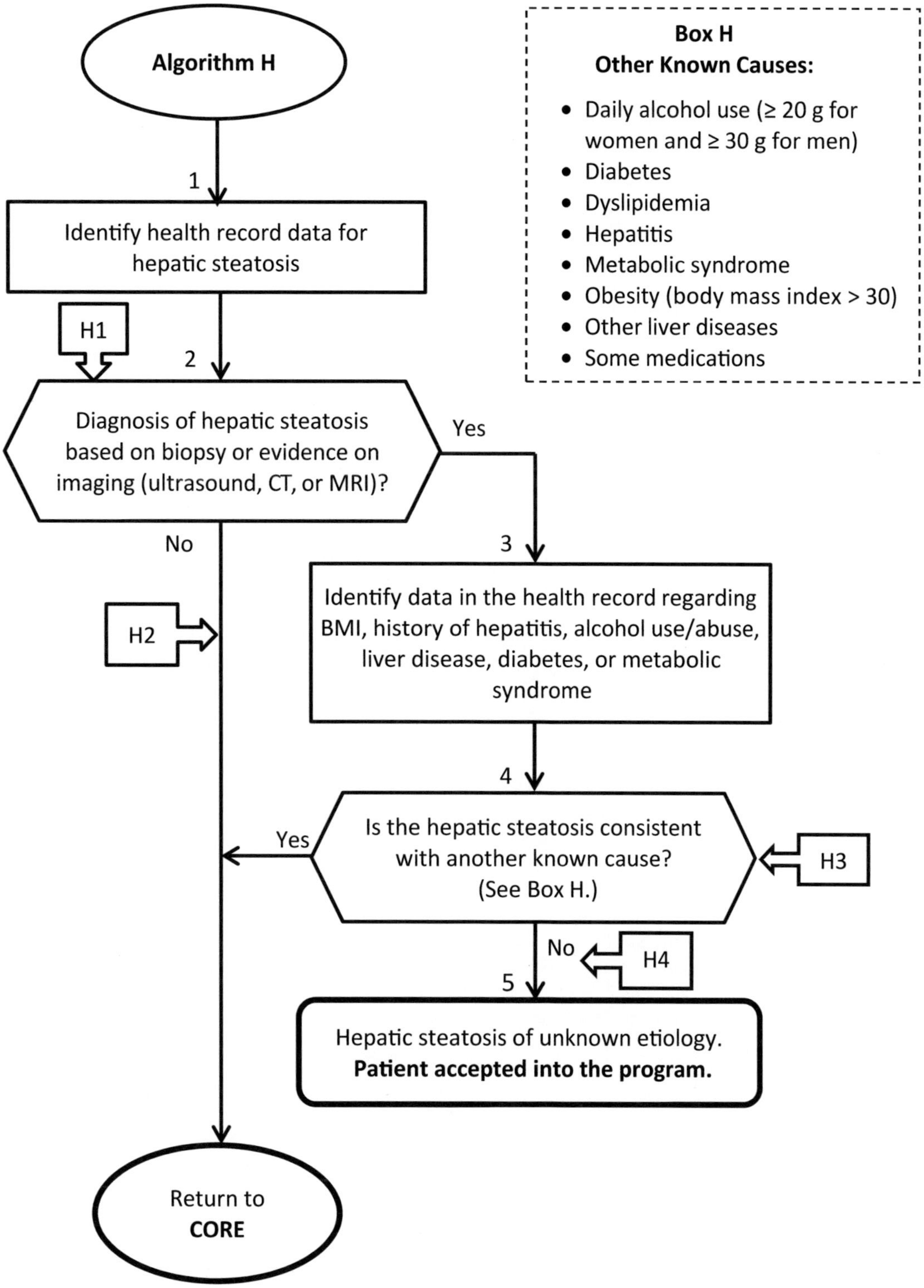

FIGURE 4-2 Revised algorithm H—Hepatic steatosis.

REFERENCES

Adams, L. A., P. Angulo, and K. D. Lindor. 2005. Nonalcoholic fatty liver disease. *Canadian Medical Association Journal* 172(7):899-905.

Adolfsson, A. 2011. Meta-analysis to obtain a scale of psychological reaction after perinatal loss: Focus on miscarriage. *Psychology Research Behavior Management* 4:29-39.

Ammon Avalos, L., C. Galindo, and D. K. Li. 2012. A systematic review to calculate background miscarriage rates using life table analysis. *Birth Defects Research Part A Clinical Molecular Teratology* 94(6):417-423.

Angulo, P. 2002. Nonalcoholic fatty liver disease. *New England Journal of Medicine* 346(16):1221-1231.

Anstee, Q. M., G. Targher, and C. P. Day. 2013. Progression of NAFLD to diabetes mellitus, cardiovascular disease or cirrhosis. *Nature Reviews, Gastroenterology and Hepatology* 10(6):330-344.

APA (American Pregnancy Association). 2014. *Miscarriage*. Available at americanpregnancy.org/pregnancycomplications/miscarriage.html (accessed July 30, 2014).

Aschengrau, A., J. M. Weinberg, L. G. Gallagher, M. R. Winter, V. M. Vieira, T. F. Webster, and D. M. Ozonoff. 2009. Exposure to tetrachloroethylene-contaminated drinking water and the risk of pregnancy loss. *Water Quality Exposure and Health* 1(1):23-34.

ATSDR (Agency for Toxic Substances and Disease Registry). 2013. *Addendum to the Toxicological Profile for Trichloroethylene*. Atlanta, GA: Agency for Toxic Substances and Disease Registry, Division of Toxicology and Human Health Sciences. Available at www.atsdr.cdc.gov/toxprofiles/tce_addendum.pdf (accessed January 2, 2015).

Attarchi, M. S., M. Ashouri, Y. Labbafinejad, and S. Mohammadi. 2012. Assessment of time to pregnancy and spontaneous abortion status following occupational exposure to organic solvents mixture. *International Archives of Occupation and Environmental Health* 85(3):295-303.

Barragan-Martinez, C., C. A. Speck-Hernandez, G. Montoya-Ortiz, R. D. Mantilla, J. M. Anaya, and A. Rojas-Villarraga. 2012. Organic solvents as risk factor for autoimmune diseases: A systematic review and meta-analysis. *Public Library of Science One* 7(12).

Bayard, M., J. Holt, and E. Boroughs. 2006. Nonalcoholic fatty liver disease. *American Family Physician* 73(11):1961-1968.

Beutel, M., R. Deckardt, M. von Rad, and H. Weiner. 1995. Grief and depression after miscarriage: Their separation, antecedents, and course. *Psychosomatic Medicine* 57(6):517-526.

Bhatia, L. S., N. P. Curzen, and C. D. Byrne. 2012a. Nonalcoholic fatty liver disease and vascular risk. *Current Opinions in Cardiology* 27(4):420-428.

Bhatia, L. S., N. P. Curzen, P. C. Calder, and C. D. Byrne. 2012b. Non-alcoholic fatty liver disease: A new and important cardiovascular risk factor? *European Heart Journal* 33(10):1190-1200.

Blackmore, E. R., D. Cote-Arsenault, W. Tang, V. Glover, J. Evans, J. Golding, and T. G. O'Connor. 2011. Previous prenatal loss as a predictor of perinatal depression and anxiety. *British Journal of Psychiatry* 198(5):373-378.

Bove, F. J., P. Z. Ruckart, M. Maslia, and T. C. Larson. 2014a. Evaluation of mortality among marines and Navy personnel exposed to contaminated drinking water at USMC Base Camp Lejeune: A retrospective cohort study. *Environmental Health* 13:10.

Bove, F. J., P. Z. Ruckart, M. Maslia, and T. C. Larson. 2014b. Mortality study of civilian employees exposed to contaminated drinking water at USMC Base Camp Lejeune: A retrospective cohort study. *Environmental Health* 13:68.

Broen, A. N., T. Moum, A. S. Bodtker, and O. Ekeberg. 2004. Psychological impact on women of miscarriage versus induced abortion: A 2-year follow-up study. *Psychosomatic Medicine* 66(2):265-271.

Christensen, K. Y., D. Vizcaya, H. Richardson, J. Lavoue, K. Aronson, and J. Siemiatycki. 2013. Risk of selected cancers due to occupational exposure to chlorinated solvents in a case-control study in Montreal. *Journal of Occupational and Environmental Medicine* 55(2):198-208.

Cooper, G. S., S. L. Makris, P. J. Nietert, and J. Jinot. 2009. Evidence of autoimmune-related effects of trichloroethylene exposure from studies in mice and humans. *Environmental Health Perspective* 117(5):696-702.

Day, C. P. 2006. Non-alcoholic fatty liver disease: Current concepts and management strategies. *Clinical Medicine* 6(1):19-25.

de Gonzalez, A. B., A. I. Apostoaei, L. H. S. Veiga, P. Rajaraman, B. A. Thomas, F. O. Hoffman, E. Gilbert, and C. Land. 2012. RadRAT: A radiation risk assessment tool for lifetime cancer risk projection. *Journal of Radiological Protection* 32(3):205-222.

Du, C. L., and J. D. Wang. 1998. Increased morbidity odds ratio of primary liver cancer and cirrhosis of the liver among vinyl chloride monomer workers. *Occupational and Environmental Medicine* 55(8):528-532.

Dzubow, R. B., S. Makris, C. S. Scott, and S. Barone, Jr. 2010. Early lifestage exposure and potential developmental susceptibility to tetrachloroethylene. *Birth Defects Research Part B—Developmental and Reproductive Toxicology* 89(1):50-65.

Eisenberg, E., and K. Brumbaugh. 2009. *Frequently asked questions: Infertility*. Department of Health and Human Services. Available at www.womenshealth.gov/publications/our-publications/fact-sheet/infertility.pdf (accessed January 22, 2015).

EPA (U.S. Environmental Protection Agency). 2011. *Toxicological review of tricholorethylene (CAS No. 79-01-6) in support of summary information on the Integrated Risk Information System (IRIS)*. EPA/635/R-08/011F. U.S. Environmental Protection Agency.

EPA. 2012. *Toxicological review of tetracholorethylene (perchloroethylene) (CAS No. 127-18-4) in support of summary information on the Integrated Risk Information System (IRIS)*. EPA/635/R-08/011F. U.S. Environmental Protection Agency.

Frost, M., and J. T. Condon. 1996. The psychological sequelae of miscarriage: A critical review of the literature. *Australian and New Zealand Journal of Psychiatry* 30(1):54-62.

Gift, T., and O. Spence. 2014. Abstract: *Infertility and Quality of Life: Findings from a Literature Review.* Paper read at CDC STD Prevention Conference, June 9-12, 2014, Atlanta, GA.

Gilbert, K. M. 2010. Xenobiotic exposure and autoimmune hepatitis. *Hepatitis Research and Treatment* 248157.

Greil, A. L. 1997. Infertility and psychological distress: A critical review of the literature. *Social Science & Medicine* 45(11):1679-1704.

Hansen, J., M. Sallmen, A. I. Selden, A. Anttila, E. Pukkala, K. Andersson, I. L. Bryngelsson, O. Raaschou-Nielsen, J. H. Olsen, and J. K. McLaughlin. 2013. Risk of cancer among workers exposed to trichloroethylene: Analysis of three Nordic cohort studies. *Journal of the National Cancer Institute* 105(12):869-877.

Howard, J. 2013. *Minimum Latency & Types or Categories of Cancer.* World Trade Center Health Program, Centers for Disease Control and Prevention.

IARC (International Agency for Research on Cancer). 2014a. Tetrachloroethylene. In *IARC Monographs on the Evaluation of Carcinogenic Risks to Humans, Vol. 106*. Lyon, France: International Agency for Research on Cancer.

IARC. 2014b. Trichloroethylene. In *IARC Monographs on the Evaluation of Carcinogenic Risks to Humans, Vol. 106*. Lyon, France: International Agency for Research on Cancer.

IOM (Institute of Medicine). 2003. *Gulf War and health, Volume 2: Insecticides and solvents*. Washington, DC: The National Academies Press.

ISN (International Scleroderma Network). Undated. *Systemic Sclerosis: CREST Syndrome*. Available at www.sclero.org/medical/about-sd/types/systemic/limited/crest/a-to-z.html (accessed September 17, 2014).

Jiang, C. M., C. W. Pu, Y. H. Hou, Z. Chen, M. Alanazy, and L. Hebbard. 2014. Non alcoholic steatohepatitis a precursor for hepatocellular carcinoma development. *World Journal of Gastroenterology* 20(44):16464-16473.

Kaiser, J. P., J. C. Lipscomb, and S. C. Wesselkamper. 2012. Putative mechanisms of environmental chemical-induced steatosis. *International Journal of Toxicology* 31(6):551-563.

Kumar, S. 2011. Occupational, environmental and lifestyle factors associated with spontaneous abortion. *Reproductive Sciences* 18(10):915-930.

Lawrence, R. C., C. G. Helmick, F. C. Arnett, R. A. Deyo, D. T. Felson, E. H. Giannini, S. P. Heyse, R. Hirsch, M. C. Hochberg, G. G. Hunder, M. H. Liang, S. R. Pillemer, V. D. Steen, and F. Wolfe. 1998. Estimates of the prevalence of arthritis and selected musculoskeletal disorders in the United States. *Arthritis and Rheumatism* 41(5):778-799.

Lee, D. 2014. *Drug induced liver disease: Steatosis (fatty liver)*. Available at www.medicinenet.com/drug_induced_liver_disease/page7.htm#steatosis_fatty_liver (accessed August 13, 2014).

Lipworth, L., J. S. Sonderman, M. T. Mumma, R. E. Tarone, D. E. Marano, J. D. Boice, Jr., and J. K. McLaughlin. 2011. Cancer mortality among aircraft manufacturing workers: An extended follow-up. *Journal of Occupational and Environmental Medicine* 53(9):992-1007.

Mascarenhas, M. N., S. R. Flaxman, T. Boerma, S. Vanderpoel, and G. A. Stevens. 2012. National, regional, and global trends in infertility prevalence since 1990: A systematic analysis of 277 health surveys. *PLoS Medicine* 9(12):e1001356.

Mayo Clinic. 2013. *Scleroderma*. Available at www.mayoclinic.org/diseases-conditions/scleroderma/basics/causes/con-20021378 (accessed August 4, 2014).

Mayo Clinic. 2014. *Infertility basics*. Available at www.mayoclinic.org/diseases-conditions/infertility/basics/causes/con-20034770 (accessed July 30, 2014).

McAvoy, N. C., J. W. Ferguson, I. W. Campbell, and P. C. Hayes. 2006. Review: Non-alcoholic fatty liver disease: Natural history, pathogenesis and treatment. *British Journal of Diabetes & Vascular Disease* 6:251-260.

Miller, F. W., L. Alfredsson, K. H. Costenbader, D. L. Kamen, L. M. Nelson, J. M. Norris, and A. J. De Roos. 2012. Epidemiology of environmental exposures and human autoimmune diseases: Findings from a National Institute of Environmental Health Sciences Expert Panel Workshop. *Journal of Autoimmunity* 39(4):259-271.

Mohanty, S. R. 2006. *Non-alcoholic fatty liver disease (NAFLD) (fatty liver)*. University of Chicago. Available at www.liverfoundation.org/downloads/alf_download_194.pdf (accessed July 28, 2014).

Mora, G. F. 2009. Systemic sclerosis: Environmental factors. *Journal of Rheumatology* 36(11):2383-2396.

Mori, Y., V. M. Kahari, and J. Varga. 2002. Scleroderma-like cutaneous syndromes. *Current Rheumatology Reports* 4(2):113-122.

NORD. 2014. *Scleroderma.* Available at www.rarediseases.org/rare-disease-information/rare-diseases/byID/69/viewFullReport (accessed August 4, 2014).

NRC (National Research Council). 2009. *Contaminated water supplies at Camp Lejeune: Assessing potential health effects.* Washington, DC: The National Academies Press.

Peters, C., M. Harling, M. Dulon, A. Schablon, J. Torres Costa, and A. Nienhaus. 2010. Fertility disorders and pregnancy complications in hairdressers—A systematic review. *Journal of Occupational Medicine and Toxicology* 5:24.

Pollard, K. M. 2012. Gender differences in autoimmunity associated with exposure to environmental factors. *Journal of Autoimmunity* 38(2-3):J177-J186.

Pollard, K. M., P. Hultman, and D. H. Kono. 2010. Toxicology of autoimmune diseases. *Chemical Research in Toxicology* 23(3):455-466.

Ruckart, P. Z., F. J. Bove, and M. Maslia. 2013. Evaluation of exposure to contaminated drinking water and specific birth defects and childhood cancers at Marine Corps Base Camp Lejeune, North Carolina: A case-control study. *Environmental Health* 12:104.

Sanyal, A. J. 2002. AGA technical review on nonalcoholic fatty liver disease. *Gastroenterology* 123(5):1705-1725.

Scleroderma Foundation. 2014. *What Is Scleroderma?* Available at www.scleroderma.org/site/PageNavigator/patients_whatis.html#.U-DiMGNY8RF (accessed August 4, 2014).

Scott, C. S., and J. Jinot. 2011. Trichloroethylene and cancer: systematic and quantitative review of epidemiologic evidence for identifying hazards. *International Journal of Environmental Research and Public Health* 8(11):4238-4272.

Storck, S. 2012. *Miscarriage.* Available at www.nlm.nih.gov/medlineplus/ency/article/001488.htm (accessed September 17, 2014).

Thoma, M. E., A. C. McLain, J. F. Louis, R. B. King, A. C. Trumble, R. Sundaram, and G. M. Buck Louis. 2013. Prevalence of infertility in the United States as estimated by the current duration approach and a traditional constructed approach. *Fertility and Sterility* 99(5):1324-1331.

van den Hoogen, F., D. Khanna, J. Fransen, S. R. Johnson, M. Baron, A. Tyndall, M. Matucci-Cerinic, R. P. Naden, T. A. Medsger, Jr., P. E. Carreira, G. Riemekasten, P. J. Clements, C. P. Denton, O. Distler, Y. Allanore, D. E. Furst, A. Gabrielli, M. D. Mayes, J. M. van Laar, J. R. Seibold, L. Czirjak, V. D. Steen, M. Inanc, O. Kowal-Bielecka, U. Muller-Ladner, G. Valentini, D. J. Veale, M. C. Vonk, U. A. Walker, L. Chung, D. H. Collier, M. E. Csuka, B. J. Fessler, S. Guiducci, A. Herrick, V. M. Hsu, S. Jimenez, B. Kahaleh, P. A. Merkel, S. Sierakowski, R. M. Silver, R. W. Simms, J. Varga, and J. E. Pope. 2013. 2013 classification criteria for systemic sclerosis: An American College of Rheumatology/European League against Rheumatism collaborative initiative. *Arthritis and Rheumatology* 65(11):2737-2747.

Varga, J. 2014. Scleroderma: Diagnosis and differential diagnosis of systemic sclerosis (scleroderma) in adults. Available at http://www.uptodate.com/contents/diagnosis-and-differential-diagnosis-of-systemic-sclerosis-scleroderma-in-adults (accessed August 13, 2014).

Viragh, E. J., K. A. Viragh, and J. Laczka. 2014. Abstract: *Prevalence of spontaneous abortion in workers in the wood-processing industry.* Paper read at Challenges for Occupational Epidemiology in the 21st Century EPICOH, June 24–27, 2014, Chicago, IL.

Wahlang, B., J. I. Beier, H. B. Clair, H. J. Bellis-Jones, K. C. Falkner, C. J. McClain, and M. C. Cave. 2013. Toxicant-associated steatohepatitis. *Toxicologic Pathology* 41(2):343-360.

Walters, T. 2014a. Presentation and Charge to Committee: Institute of Medicine Committee to Define Clinical Terms in P.L. 112-154 & Review Clinical Guidelines. Washington, DC, May 15, 2014.

Walters, T. 2014b. Questions posed by IOM committee and subsequent answers from Dr. Terry Walters, U.S. Department of Veterans Affairs, July 8, 2014.

Weinhold, B. 2009. A clearer view of TCE: Evidence supports autoimmune link. *Environmental Health Perspective* 117(5):A210.

WGO (World Gastroenterology Organisation). 2012. *Nonalcoholic fatty liver disease and nonalcoholic steatohepatitis. World Gastroenterology Organization Global Guidelines.* Available at www.worldgastroenterology.org/assets/export/userfiles/2012_NASH%20and%20NAFLD_Final_long.pdf (accessed August 13, 2014).

5

Use of the Guidance

The U.S. Department of Veterans Affairs (VA) created the Camp Lejeune Health Program in response to the 2012 Janey Ensminger Act. To implement the program, VA has developed guidance to help clinicians and other health care providers make decisions about whether veterans and family members who are administratively eligible[1] for the program, have a medical condition that is covered by the act. Should a qualifying medical condition be found, veterans have their co-payments waived for treatments of the covered condition, and VA will reimburse family members as the last payer for private sector health care related to the conditions. The VA guidance was developed with input from specialists in the conditions that are covered by the act (e.g., neurology, gastroenterology, oncology, nephrology, gynecology, and mental health) (Walters 2014a,b).

VA asked the committee to assess the scientific soundness of the guidance for the designated health outcomes. The committee was not asked to comment on the broader issues of the implementation of, administration of, training for, or evaluation of the Camp Lejeune Health Program itself. While the broader context of the overall Camp Lejeune Health Program is important to the success of the guidance, assessing the entire program was not part of the committee's charge. The guidance materials are based not only on scientific evidence, but also on existing VA policies and congressional mandates in the legislation. In part, this is because there is a lack of definitive scientific evidence on which to base some decisions such as the role of environmental contaminants in the development of cancer. The act itself states that hospital care and medical services are to be provided "notwithstanding that there are insufficient medical evidence to conclude that such illnesses or conditions are attributable to such service" at Camp Lejeune. There are also various precedents for the coverage of treatment for veterans who have been exposed to toxicants, such as those established by the Agent Orange Act of 1991 (P.L. 102-4) and Gulf War legislation (Veterans Health Care Act of 1992, P.L. 102-585). Such policy decisions can help reduce the administrative burden that Camp Lejeune veterans and family members—and their health care providers—face in proving an association between residing at Camp Lejeune during the period of drinking water contamination and the concurrent or subsequent development of a covered health condition.

Users of the guidance may include not only VA clinicians who treat veterans, but also community providers and others who treat veterans and their family members. The users may also include VA financial services center personnel who review claims for reimbursement for treatment costs from veterans and their family members. This is the first time that VA has extended benefits to family members. Thus, the program presents VA with adminis-

[1] Lived at Camp Lejeune for at least 30 days between January 1, 1957, and December 31, 1987.

trative and outreach challenges, including attempting to locate the Marine Corps veterans and their families who resided at Camp Lejeune more than 25 years ago.

In this chapter, the committee considers the utility of the guidance for clinicians and discusses those aspects of the guidance for which clarification may be helpful or where the presentation might be improved. The committee's ability to offer comments was limited by the fact that the guidance is in draft form and has not been widely disseminated or used. The committee did not attend any training sessions for clinicians on using the guidance nor did it hear from any clinicians who have experience following the guidance with their patients. VA has stated that changes to the guidance are expected in the future as it is implemented throughout the department and across the country. The full text of the guidance and the algorithms are available in Appendix B.

PURPOSE OF THE GUIDANCE AND ALGORITHMS

The stated purposes of the guidance (page 1) are to (1) help a health care provider determine if a veteran or family member has a condition that is covered by the Camp Lejeune legislation, and (2) determine if an episode of care is related to the covered condition. The majority of the guidance focuses on assisting health care providers—VA, purchased care, or community clinicians—to determine if the veteran or family member has a condition that is covered by the Camp Lejeune legislation. In particular, the algorithms that accompany the guidance, both the core algorithm and the four condition-specific ones, are easy-to-follow tools that can be used to help determine whether the veteran or family member has the condition or whether alternative causes may preclude coverage for the condition. In the prior chapters, the committee has made specific recommendations for improving the algorithms for the covered conditions.

The second purpose of the guidance—to determine if a treatment or service is associated with a covered condition—is not discussed substantively in the guidance with the exception of the three bullets on page 6 that describe the extent of comprehensive coverage during active cancer treatment and the reimbursement of family members for primary and secondary conditions. Because treatment of the covered conditions is highly individualized and the specific treatments for covered conditions are not further elaborated upon in the guidance, it may be most appropriate to delete this stated purpose on page 1. This is because a clinician or family members may expect that the guidance can help them determine if a particular treatment will be covered. More discussion of treating covered conditions is included in the following section on decision points.

In the Background section of the guidance (page 3), there is information on the extent of VA coverage of hospital care and medical services—including screening procedures—pertaining to the conditions in the legislation. The guidance specifies that VA will reimburse eligible family members for screenings related to the 15 covered conditions if clinically indicated or if recommended by the U.S. Preventive Services Task Force only if the outcome of that screening leads to the diagnosis of a covered condition.

Clinically indicated screenings for Camp Lejeune veterans enrolled in VA health care are included in their comprehensive health benefits and thus do not require a co-pay. The committee finds that the diagnosis of a covered condition may require screening as well as a diagnostic evaluation at the discretion of the clinician, but the guidance does not indicate whether a diagnostic evaluation will be covered.

The committee recommends that VA revise the sentence on page 3 of the guidance to read "VA will reimburse eligible family members for screening and diagnostic evaluations that are clinically indicated, or recommended by the U.S. Preventive Services Task Force, and that lead to a diagnosis of a covered condition."

DECISION POINTS

As described in Chapter 1, the guidance uses three decision points to assess whether an illness, injury, or medical condition is eligible for coverage under the Camp Lejeune program. The decision points are incorporated into the algorithms in the guidance for each covered condition. In the sections below, the committee considers the usefulness of the decision points and indicates where improvements to the guidance and the algorithms might increase clarity and where inconsistencies could be corrected.

(1) Does the Camp Lejeune program participant have one or more of the covered conditions?

The committee considered three topics for this decision point: referrals, secondary conditions, and symptom onset and duration. These topics are discussed below.

Referrals

The committee expects that although many of the health conditions in the Janey Ensminger Act are familiar to primary care physicians, internists, family practitioners, and other health care professionals—and may be indicated by some screening and diagnostic evaluations—a specialist may be required for the diagnosis of some of those conditions (including differential diagnosis) and for treatment. Such specialists might include an oncologist for cancer; a nephrologist for some kidney diseases; a psychologist, psychiatrist, developmental pediatrician, or neurologist (including experts in substance use) from neurobehavioral effects; and a rheumatologist for scleroderma.

The guidance does not indicate when referrals to specialists should be made, nor who would review any medical records to see if a diagnosis for inclusion in the Camp Lejeune program was correct.

The committee recommends that referrals to specialists should be made when clinically indicated to obtain a definitive diagnosis and that VA should have a standardized process for making such referrals.

Secondary Conditions

The guidance states that VA has the authority to reimburse family members for medical conditions that are secondary to a covered condition. The committee finds that although the algorithm for female health in the guidance acknowledges that medical complications may ensue following female infertility or miscarriage as a result of residing at Camp Lejeune, no further guidance is given for these conditions. Furthermore, there is no acknowledgment in the descriptions or algorithms for the other covered conditions that secondary conditions and medical complications can result not only from the presence of the condition itself, but also from disease progression and from treatment for the condition. Examples include hepatic steatosis progressing to cirrhosis, a miscarriage or infertility that result in prolonged depression, or a treatment for breast cancer leading to lymphedema that requires treatment even if the cancer is in remission.

The committee recommends that VA consider adding the need to diagnose and treat secondary conditions to the descriptions or algorithms for the covered primary conditions.

Symptom Onset and Duration

In general the committee notes that, where specified, the determinations of the time of onset and duration of the covered conditions are appropriate. However, the time of onset and duration are not specified for every condition and can vary. For example, miscarriage or infertility is expected to occur during residence at Camp Lejeune but not after exposure has ceased, while a cancer will generally not occur for many years after exposure and may not exist or be evident at the time of exposure or for years afterward. In addition, some information, such as other possible causes or diagnostic criteria, is not reported consistently for each outcome. VA has made a policy decision that the time of onset matters for miscarriage and infertility but is not a consideration for cancer. This variability in criteria for each outcome may result in confusion on the part of the Camp Lejeune veterans, their family members, and clinicians. The committee has proposed a table to capture these domains for each outcome (see Table 5-1). This

TABLE 5-1 Criteria for Onset, Duration, and Exclusions for All Covered Conditions

	Neurobehavioral Effects			
	Neurobehavioral Adult Exposure	Neurobehavioral Childhood or In Utero Exposure	Parkinson's Disease	Renal Toxicity
Description/ criteria	Chronic symptoms, including Delayed reaction times Problems with short-term memory, visual perception, attention, color vision	Drug addiction Bipolar depression Neurological problems associated with neural tube defects	Parkinson's Disease Foundation or other accepted criteria	Chronic kidney disease eGFR < 60 or protein urea or kidney biopsy
Onset	During CL residence	Unknown	After CL residence	After CL residence
Duration	Persistent or intermittent since residence at CL	Unknown	n/a	n/a
Other likely causes	Alzheimer's disease Dementia ALS ADHD Basal ganglia disorders Bipolar disorder Multiple sclerosis Parkinson's disease Reductions in color discrimination, hearing, or olfactory functions Schizophrenia PTSD OCD Panic disorder			Diabetes Hypertension Volume depletion Severe heart failure Acute tubular necrosis occurring with hypotension or nephrotoxic agents Acute interstitial nephritis due to medication Obstructive uropathy Hypertensive nephrosclerosis Sickle cell kidney disease HIV-associated nephropathy
Additional notes	None	Unknown	None	Atypical course (faster progression) may indicate exacerbation by CL exposure

NOTE: ACR = American College of Rheumatology; ADHD = attention deficit hyperactivity disorder; ALS = amyotrophic lateral sclerosis; CL = Camp Lejeune; CT = computerized tomography; eGFR = estimated glomerular filtration rate; HIV = human immunodeficiency virus; MRI = magnetic resonance imaging; n/a = not available; OCD = obsessive compulsive disorder; PTSD = posttraumatic stress disorder; red = committee additions.

This table assumes that the patient is administratively eligible for the program (served on active duty or resided at CL for not less than 30 days between January 1, 1957, and December 31, 1987).

Cancers	Scleroderma	Miscarriage	Infertility	Hepatic Steatosis
Breast cancer Bladder cancer Esophageal cancer Kidney cancer Leukemia Lung cancer Multiple myeloma Myelodysplastic syndromes Non-Hodgkin's lymphoma	ACR diagnostic criteria	Chronic persistent physical or mental health conditions associated with miscarriage	Chronic persistent physical or mental health conditions associated with infertility	Identified by ultrasound, CT, or MRI, or a biopsy
Any time during or after CL residence	Any time during or after CL residence	Miscarriage during CL residence	Infertility during CL residence	During CL residence, although may have been subclinical
n/a	n/a	Problems with miscarriage resolve during or shortly after CL residence	Problems with infertility resolve during or shortly after CL residence	n/a
Not considered	Not considered	Not considered	Not considered	Alcohol abuse Obesity Metabolic syndrome Some medications Hepatitis Dyslipidemia Other liver diseases
All medical care is covered during the duration of cancer treatment (surgery, radiation, chemotherapy, immunotherapy, and hormonal therapy) to be certified by the treating physician at 6-month intervals	None	None	None	None

table includes more information than is available in the guidance and algorithms and is based on the committee's expertise and discussion in the previous chapters.

In general, the committee finds that the guidance is correct in indicating that some of the conditions or their symptoms will be evident during the time the veteran or family member resided at Camp Lejeune and was exposed to the contaminated drinking water. This is not explicitly stated in the guidance; however, for some conditions, such as cancer, the guidance does not state that cancer may have been evident at the time of exposure, only that it can have a long latency period.

It would be helpful if the guidance included a table (such as Table 5-1) or had a standardized format for the discussion of each health outcome to allow the clinician to quickly determine the criteria for the onset of the condition, the duration of the condition, and the exclusionary factors for each condition. There could be a comment section in the table or text that would indicate any mitigating or other factors that should be considered when determining if the patient had a covered health condition. Alternatively, the annotations to the algorithms could present this information in a standardized format. In the current guidance, the format and content of the annotations for each algorithm are variable, and the annotations do not always track with the algorithm boxes that refer to them. The committee believes that when assessing a patient in real time clinicians are more likely to refer quickly to a table or algorithm than they are to read a lengthy text in the guidance.

The committee recommends the following: that VA specify details for the same domains (such as criteria for the diagnosis, the onset and duration, as well as other possible causes and exclusionary factors) for all covered conditions in order to ensure clarity, completeness, and consistency; that VA consider revising the text in the guidance on page 4, "Covered conditions whose onset occurs at the time of solvent exposure" to reflect the recommended revisions for neurobehavioral effects in adults and the new algorithm for children; and that VA consider removing neurobehavioral effects from the first sentence in this section because not all such effects may be evident *during* exposure.

(2) Is there evidence that the condition occurred as a result of a cause other than residence at Camp Lejeune?

The committee considered this decision point to be the most problematic, in part because there is some discrepancy with the original legislation and in part because the terminology used to assess covered conditions is inconsistent throughout the guidance. On page 1 in the second bullet under Key Points, the guidance states, "[H]ospital care and medical services may not be furnished . . . for an illness or condition of a Camp Lejeune Veteran or family member that is found, in accordance with guidelines issued by the Under Secretary for Health, to have resulted from a cause other than the residence at Camp Lejeune." On pages 3 and 4, the guidance states again that veterans cannot receive care for a covered condition if the condition has resulted from a cause other than residence at Camp Lejeune. However, this language is in contradiction to both the act and the language on page 4 of the guidance in the first sentence under Decision Point #2.

The committee recommends that VA state whether veterans must meet the same criteria as family members regarding other possible causes for a condition.

The guidance also uses inconsistent terminology in assisting clinicians in determining whether the condition has another cause. For example, for neurobehavioral effects, the clinician should determine if the symptoms "are as likely as not, related to exposure to volatile organics in the past" (page 8); for renal toxicity, the clinician "should consider whether it is probable" that the kidney disease results from something other than solvent exposure, and the clinician "might reasonably conclude that the renal disease is as likely as not associated" with another cause (page 10). Finally for hepatic steatosis, the guidance also asks the clinician to "consider whether it is more likely than not" that the fatty liver disease has another cause (page 10). This mix of terminology may be confusing to both the patient and the clinician, and no assistance is given on how to determine what is "more likely than not," particularly in light of the following statement, which appears early in the guidance: "In cases where there is reasonable doubt as to the diagnosis or primary cause for the diagnosis, clinicians should resolve in favor of the Camp Lejeune Veteran or family member" (page 2). Although the text accompanying the algorithms provides some information on what is meant by "consistent," VA may want to consider providing more information for clinicians as the program evolves. Furthermore, veterans and family members with cancer or scleroderma are not assessed for other possible causes for their disease and clinicians are not required to rule out other causes, such as smoking as a possible cause of lung cancer.

The committee finds that the language in the guidance is inconsistent with regard to the level of association necessary to link exposure to drinking water at Camp Lejeune with a covered condition.

The committee recommends that VA set one standard for the likelihood that a condition (with the exception of cancer and scleroderma) must be related to residence at Camp Lejeune. The committee also recommends that VA reword the decision point to read "Is there evidence that the condition is as likely as not to have occurred as a result of a cause other than residence at Camp Lejeune?" in order to more accurately reflect the rest of the guidance.

(3) Is the episode of care or treatment related to the covered condition?

In several instances the guidance asks clinicians to "verify" or "certify" information pertaining to whether or not a specific visit, treatment, or secondary condition is related to a covered condition ("Certify" appears on page 6 in bullets 1 and 3; "verify" appears on page 5 in the first paragraph in Decision Point #3). When the committee asked for clarification of these terms, VA indicated that it expects clinicians to document whether the encounter or treatment is related to a condition in their note for billing purposes (Walters, 2014a). In the case of cancer, VA intends to ask that the clinician "certify" the duration of treatment. While this may be clear to VA clinicians, it may not be evident to non-VA clinicians who treat family members. There is no further information in the guidance on record keeping or on how a clinician should "verify" or "certify" pertinent information.

The committee finds that the guidance is unclear regarding what health care providers must do in order to certify or verify that a treatment or service is provided for one of the covered conditions and what documentation must be submitted, particularly by non-VA health care providers in order to ensure the treatment is covered (and that there is no co-pay for veterans). It would be helpful if instructions on providing this information was electronically available, e.g., on the VA Camp Lejeune website, so that both participants and their health care providers could access it. It would be useful if such information were included in the clinical guidance or, at the very least, if the clinical guidance contained a reference or link to where more information could be obtained.

The committee recommends that VA include instructions to clinicians about how to record essential information regarding their patients' diagnoses and treatments for those conditions.

FUTURE CONSIDERATIONS

The committee expects—and VA has indicated—that the guidance will be revised as VA receives feedback from clinicians on its utility and clarity, and as additional scientific information becomes available pertaining to the covered outcomes and their relationship to drinking water contaminants. This may be of particular importance for the neurobehavioral effects and renal toxicity endpoints as epidemiologic and toxicologic research continues on the association between exposure to the drinking water contaminants found at Camp Lejeune and adverse health effects in those two domains. As new research is published, VA may want to consider a process to evaluate periodically new research on all endpoints and to revise the guidance accordingly to ensure that it fulfills its intended purpose—as is now done for the VA/U.S. Department of Defense Clinical Practice Guidelines for various health conditions.

The committee recognizes that the VA guidance can only address the illnesses and medical conditions listed in the Janey Ensminger Act. However, unlike the Agent Orange and Gulf War legislation, the Camp Lejeune legislation does not allow VA to incorporate new scientific evidence that may indicate new associations between

Camp Lejeune exposure and adverse health conditions or to revise the ones listed in the legislation. Even the act recognizes "that there is insufficient medical evidence to conclude that such illnesses or conditions are attributable to such service." Future information may show clear links with diseases not currently listed in the act, may provide convincing evidence that health outcomes currently associated with contaminated drinking water are spurious or explained by other factors, or may more clearly define specific neurobehavioral domains and kidney pathology affected by those exposures.

Important information sources include new studies and meta-analyses by authoritative bodies such as the International Agency for Cancer Research (IARC) and the U.S. Environmental Protection Agency. For example, in 2014, IARC released updated assessments of the carcinogenicity of trichloroethylene (TCE) and perchloroethylene (PCE) that found there was limited evidence of an association between exposure to PCE and bladder cancer, and that there was sufficient evidence in animals and humans of an association between exposure to TCE and both non-Hodgkin's lymphoma and liver cancer (IARC, 2014). Future research may provide new associations with other critical health conditions such as those in children where an emerging literature suggests that solvent exposure may be linked with posttraumatic stress disorder, schizophrenia, and other neurobehavioral effects (Aschengrau et al., 2012).

The committee appreciates that much of the clinical guidance is the result of VA policy decisions and interpretations of congressional intent in the legislation. The committee also understands that the guidance was not developed based solely on scientific evidence (e.g., acceptance of all specified cancers without regard for their latency or for possible contributing factors such as smoking), and in fact this "insufficient medical evidence to conclude that such illnesses or conditions are attributable to such service" is specifically stated in the legislation. The committee agrees with the VA guidance for clinical conditions with poorly defined diagnostic criteria that states "In cases where there is reasonable doubt as to the diagnosis or primary cause for the diagnosis, clinicians should resolve in favor of the Camp Lejeune Veteran or family member" (page 5). In its assessment of the clinical guidance and the scientific evidence used to characterize renal toxicity and neurobehavioral effects, the committee has tried to give the benefit of the doubt to the veteran and family members, particularly when expert judgment was required.

VA has done a commendable job in dealing with a scientifically and administratively complex task. The committee hopes that the above recommendations in this report will clarify and enhance the guidance document so that Camp Lejeune veterans and their family members can receive hospital care and medical services under the Janey Ensminger Act with a minimum of confusion for them and for the clinicians from whom they seek care.

REFERENCES

Aschengrau, A., J. M. Weinberg, P. A. Janulewicz, M. E. Romano, L. G. Gallagher, M. R. Winter, B. R. Martin, V. M. Vieira, T. F. Webster, R. F. White, and D. M. Ozonoff. 2012. Occurrence of mental illness following prenatal and early childhood exposure to tetrachloroethylene (PCE)-contaminated drinking water: A retrospective cohort study. *Environmental Health* 11:2.

IARC (International Agency for Research on Cancer). 2014. Trichloroethylene. In *IARC Monographs on the Evaluation of Carcinogenic Risks to Humans, Vol. 106*. Lyon, France: International Agency for Research on Cancer.

Walters, T. 2014a. Presentation and Charge to Committee: Institute of Medicine Committee to Define Clinical Terms in P.L. 112-154 & Review Clinical Guidelines. Washington, DC, May 15, 2014.

Walters, T. 2014b. Questions posed by IOM committee and subsequent answers from Dr. Terry Walters, U.S. Department of Veterans Affairs, July 8, 2014.

Appendix A

Committee Biographies

David R. Nerenz, Ph.D. (*Chair*) is the director of the Center for Health Policy and Health Services Research at the Henry Ford Health System in Detroit, Michigan, and is also the director of outcomes research at the Neuroscience Institute and vice chair for research in the Department of Neurosurgery at Henry Ford Hospital. He has served on the National Committee for Quality Assurance's Culturally and Linguistically Appropriate Services Workgroup and is currently a commissioner on the Medicare Payment Advisory Commission. Dr. Nerenz has served in various roles with the Institute of Medicine, including as chair of the Committee on Leading Health Indicators for Healthy People, 2020, and as a committee member on Gulf War Veterans: Treating Symptoms and Syndromes and on Gulf War Veterans: Measuring Health. He serves on the editorial boards of *Population Health Management* and *Medical Care Research and Review*. Dr. Nerenz received a Ph.D. in social psychology from the University of Wisconsin–Madison.

Robert J. Alpern, M.D., is the Ensign Professor of Medicine and dean at Yale University Medical School. Previously, he served as the chief of the Division of Nephrology at the University of Texas Southwestern Medical Center, where he also held the Ruth W. and Milton P. Levy, Sr., Chair in Molecular Nephrology and the Atticus James Gill, M.D., Chair in Medical Science. In July 1998, Dr. Alpern was appointed dean of Southwestern Medical School, a position that he held until 2004. Dr. Alpern's research has focused on the regulation of kidney transport proteins. In addition, Dr. Alpern has been highly committed to teaching and clinical medicine. He was elected president of the American Society of Nephrology in 2000, was elected to the American Society of Clinical Investigation and the Association of American Physicians, and has served on the advisory council of the National Institute of Diabetes and Digestive and Kidney Diseases. He was elected to the Institute of Medicine in 2007. Dr. Alpern received his M.D. degree from the University of Chicago.

Paolo Boffetta, M.D., M.P.H., is the director of the Institute for Translational Epidemiology and the associate director for population sciences of The Tisch Cancer Institute at Mount Sinai Hospital in New York. He holds additional appointments in the Department of Epidemiology at the Harvard School of Public Health, the Department of Biomedical Sciences and Human Oncology at the University of Turin (Italy), the Department of Medicine at Vanderbilt University, and numerous other international academic and health institutions. Dr. Boffetta's research interests focus on molecular and genetic epidemiology, cancer epidemiology, cardiovascular and diabetes epidemiology, cancer prevention, and epidemiologic methods. He is the author of more than 970 peer-reviewed publications, and he has edited 12 books and authored more than 80 book chapters, mainly in the field of the epidemiology of chronic diseases. He is an associate editor of *Annals of Oncology*, *European Journal of Clinical Investigation*, and *Frontiers in Cancer Epidemiology* and is a member of more than 10 editorial boards of scientific journals. He is the founding member of several international cancer epidemiology consortia and networks. Dr. Boffetta completed his M.D. at the University of Turin.

Mary E. Davis, Ph.D., is a professor in the Department of Physiology and Pharmacology of the West Virginia University Health Sciences Center. Her research interests are in the toxicology of environmental and occupational pollutants, including water-disinfection byproducts, halogenated solvents, and arsenic. She is particularly interested in the mechanisms of toxicity in the liver, kidneys, and vascular system. Dr. Davis was treasurer of the Society of Toxicology and is a former president of the society's Allegheny–Erie Regional Chapter. She has served on the U.S. Environmental Protection Agency Science Advisory Board and the editorial boards of *Toxicology* and *Toxicology and Applied Pharmacology*. She was a member of the National Research Council Committee on Assessing Human Health Risks of Trichloroethylene and the Committee on Tetrachloroethylene. She received her Ph.D. in pharmacology from Michigan State University.

Michael E. Goldberg, M.D., is the David Mahoney Professor of Brain and Behavior in the Department of Neuroscience at Columbia University College of Physicians and Surgeons. His research is focused on the physiology of cognitive processes: visual attention, spatial perception, and decision making. He is a past president of the Society for Neuroscience and was elected to the National Academy of Sciences (NAS) in 2011. He is currently serving on the NAS Temporary Nominating Group for Class II: Biological Sciences. Dr. Goldberg received his M.D. from Harvard Medical School. He is an active clinical neurologist, serving on the Columbia University Medical Center Neurology Hospitalist team.

Paul Grundy, M.D., M.P.H., FACOEM, FACPM, is IBM's global director of health care transformation. In this role, Dr. Grundy develops and executes strategies that support IBM's health care industry transformation initiatives. Part of his work is directed toward shifting health care delivery around the world toward consumer-focused, primary-care based systems through the adoption of new philosophies, primary-care pilot programs, new incentives systems, and the information technology required to implement such changes. Dr. Grundy is also the founding president of the Patient Centered Primary Care Collaborative and is an adjunct professor at the University of Utah School of Medicine, Department of Family and Preventive Medicine. An active social entrepreneur and speaker on global health care transformation, Dr. Grundy is driving comprehensive, linked, and integrated health care and the concept of the patient-centered medical home. He is a member of the Institute of Medicine and a recipient of the 2012 National Committee for Quality Assurance (NCQA) Quality Award. Dr. Grundy received his M.D. from the University of California, San Francisco, and his M.P.H. from the University of California, Berkeley.

Nancy L. Keating, M.D., M.P.H., is professor of health care policy at Harvard Medical School and an associate physician at Brigham and Women's Hospital. Dr. Keating's research focuses on the quality of care delivered to patients with cancer and the influence of physicians, hospitals, and health care systems on care delivery as well as communication between patients and physicians and among physicians. She recently served as co–principal investigator of a large evaluation of the quality of care delivered to patients diagnosed with cancer in the U.S. Department of Veterans Affairs health care system. Dr. Keating has also served as a key investigator in the Cancer Care Outcomes Research and Surveillance (CanCORS) Consortium, a long-term prospective study examining patterns of care and outcomes for patients with lung and colorectal cancer. Her research has been funded by the National Cancer Institute, the American Cancer Society, the Prostate Cancer Foundation, the Doris Duke Foundation, and the Komen for the Cure Foundation. She currently serves as a member of the editorial board of the *Journal of Geriatric Oncology*, and she is a member of the American Society of Clinical Oncology Guidelines Panel and the National Comprehensive Cancer Center Senior Oncology Guideline Panel. She received her M.D. degree from the University of Chicago, Pritzker School of Medicine, and her M.P.H. degree from the Harvard School of Public Health.

Patricia Janulewicz Lloyd, D.Sc., is an assistant professor of environmental health at the Boston University School of Public Health. Dr. Janulewicz Lloyd has conducted several studies of the effects of perchloroethylene and other solvents on the developing nervous system. These include assessments of neuropsychological functioning, vision, mental illness, risky behavior, and learning disorders. She earned both her doctorate and master's degrees from the Boston University School of Public Health.

Gary O. Rankin, Ph.D., is a professor in the Department of Pharmacology, Physiology and Toxicology at Marshall University. He joined the then-newly formed Marshall University School of Medicine in 1978 and has been the chair of the department since 1986. He also served as associate dean for biomedical graduate education and research development (1989–1992). Dr. Rankin's research interests are in the area of renal toxicology and include the nephrotoxicity induced by succinimides and aminochlorophenols, urotoxicity induced by halogenated anilines, renal transport of drug metabolites, and the bioactivation of chemicals to nephrotoxicant species. Dr. Rankin is a member of several professional societies, including the Association of Medical School Pharmacology Chairs, the Society of Toxicology (SOT), the American Society for Pharmacology and Experimental Therapeutics (ASPET), and the West Virginia Academy of Sciences. He has chaired several committees for SOT and ASPET and has held several offices in the Ohio Valley SOT Chapter, including president. He was a regular member of NIH TOX1 and ALTOX4 Study Sections (1994–1999) and serves as a frequent reviewer for NIH special study sections (SBIR; F32s; IDeA CTRs) and the newly formed PBKD study section. Dr. Rankin served as an associate editor of *Toxicology and Applied Pharmacology* (1995–2001) and has been an editorial board member for *Toxicology* since 1994. He has authored or co-authored more than 120 peer-reviewed manuscripts, 8 review articles, 17 book chapters, and more than 240 research presentations at local, regional, national, and international meetings. He earned a doctoral degree from the University of Mississippi in medicinal chemistry.

Mark J. Utell, M.D., is a professor of medicine and environmental medicine, the director of occupational and environmental medicine, and the former director of pulmonary and critical care medicine at the University of Rochester Medical Center. His research interests have centered on the effects of environmental toxicants on the human respiratory tract. Dr. Utell has published extensively on the health effects of inhaled gases, particles, and fibers in the workplace and other indoor and outdoor environments. He was the co–principal investigator of an U.S. Environmental Protection Agency (EPA) particulate matter center and is a former chair of the Health Effects Institute's research committee. He has served as chair of EPA's Environmental Health Committee and on the executive committee of the EPA Science Advisory Board. He is a former recipient of the National Institute of Environmental Health Sciences Academic Award in Environmental and Occupational Medicine. Dr. Utell is currently a member of the National Research Council's (NRC's) Committee on Strengthening the U.S. EPA Laboratory Enterprise. He previously served a chair of the Institute of Medicine (IOM) Committee to Review the Department of Labor's Site Exposure Matrix and was on the NRC Board on Environmental Studies and Toxicology; served as chair of the NRC Committee to Review the NIOSH Respiratory Disease Research Program and the Committee to Review the Department of Defense Enhanced Particulate Matter Surveillance Program Report; and served as a member of the NRC Committee on Research Priorities for Airborne Particulate Matter and Research Strategy for Environmental, Health, and Safety Aspects of Engineered Nanomaterials and the IOM Committee on Gulf War and Health: Literature Review of Selected Environmental Agents, Pollutants, and Synthetic Chemical Compounds. He received his M.D. from Tufts University School of Medicine.

Carol S. Wood, Ph.D., is a staff scientist in the Environmental Sciences Division of Oak Ridge National Laboratory. She has more than 19 years of experience as a toxicologist at Oak Ridge National Laboratory with extensive work in risk assessment for inhalation/pulmonary and oral toxicity of heavy metals and pesticides. She has worked on acute exposure guideline levels and provisional advisory levels, in which health-based exposure levels are developed for priority toxic chemicals. These projects often use toxicokinetic data and PBPK models for extrapolation from animals to humans. She serves on the board of directors for the American Board of Toxicology. At the request of U.S. Environmental Protection Agency, she wrote the guidance document "Standard Evaluation Procedure for Developmental Neurotoxicity Studies" and reviewed numerous submissions of testing and positive control neurotoxicity data. Her research experience and interests include models of developmental, reproductive, and neurotoxic outcomes from environmental contaminants. Dr. Wood is currently certified in general toxicology by the American Board of Toxicology. She earned a Ph.D. in toxicology from Oregon State University in 1993 with emphasis in developmental and reproductive toxicology.

Albert W. Wu, M.D., M.P.H., is a professor of health policy and management at the Johns Hopkins Bloomberg School of Public Health, with joint appointments in epidemiology, international health, medicine, and surgery and at the Carey Business School. His research and teaching focus on patient outcomes and quality of care. He is the director of the Johns Hopkins Center for Health Services and Outcomes Research, the Ph.D. program in health services research, and the certificate program in quality, patient safety, and outcomes research. He was a senior adviser to the World Health Organization Patient Safety Programme. He leads studies to assess patient-reported outcomes in electronic health records; to connect East Baltimore community-based organizations to one another and the health system; to prevent injurious falls in older adults; and to support "second victims"—health care workers traumatized by adverse patient events. He was a member of the Institute of Medicine committees on Preventing Medication Errors and on Readjustment Needs of Returning Service Members from the Iraq and Afghan Wars. He has more than 370 peer-reviewed publications and maintains a clinical practice in general internal medicine. Dr. Wu received his M.P.H. from the University of California, Berkeley, and his M.D. from the Cornell University School of Medicine.

Appendix B

Guidance for VHA Staff Honoring America's Veterans and Caring for Camp Lejeune Families Act of 2012, Section 102, Covered Clinical Conditions

VA HEALTH CARE | Defining EXCELLENCE in the 21st Century

U.S. Department of Veterans Affairs
Veterans Health Administration
VA Camp Lejeune Task Force

Guidance for VHA Staff
Honoring America's Veterans and Caring for Camp Lejeune Families Act of 2012, Section 102, Covered Clinical Conditions

September 2013

PURPOSE AND KEY POINTS

Purpose:

To provide guidance for clinicians who are supporting qualified Camp Lejeune Veteran or family member who are covered by Section 102 of Public Law 112-154, *Honoring America's Veterans and Caring for Camp Lejeune Families Act of 2012.* This guidance will assist healthcare providers in determining whether the Veteran or family member has a medical condition or illness that is covered by the law and will also assist in determining whether the episode of care is related to the covered condition.

According to its published rules, VA will provide care in its healthcare system facilities for eligible Veterans. Camp Lejeune Veterans who qualify will be Priority Level 6, will receive the VHA uniform benefits package, and will have their copayments waived for care related to the 15 covered conditions. VA will reimburse as the last payer eligible Camp Lejeune family members for private sector healthcare related to the 15 conditions covered by the law. Family members are not eligible for care within VA healthcare facilities.

Key Points:

- Section 102 of Public Law 112-154 addresses healthcare for eligible Veterans and family members with 15 covered conditions. It does not change existing disability compensation determinations for Veterans exposed to contaminated water at Camp Lejeune.

- The law does not give VA authority to cover conditions other than the 15 designated conditions and illnesses and "hospital care and medical services may not be furnished…for an illness or condition of a Camp Lejeune Veterans or family member that is found, in accordance with guidelines issued by the Under Secretary for Health, to have resulted from a cause other than the residence" at Camp Lejeune."

- In cases where there is reasonable doubt as to the diagnosis or primary cause for the diagnosis, clinicians should resolve in favor of the Camp Lejeune Veteran or family member.
- VA is required by law to be the last payer for treatments provided for the 15 covered conditions for family members. Clinical review using these guidelines may be needed to determine if a clinic visit or treatments were related to one of the 15 covered conditions.
- Clinical input is required for three decision points:

  1. Does the applicant have one of the medical illnesses or conditions specified in the law?
  2. Is there another cause for the medical illness or condition?
  3. Which treatments are associated with the medical illness or condition?

- There are some medical conditions in the law that do not have well established diagnostic criteria or have many potential causes. In these cases clinical judgment guided by the applicant's medical history and available scientific findings will be needed to determine if a Camp Lejeune Veteran or family member has a condition covered by the law.
-

BACKGROUND

Public Law 112-154 (subsequently referred to as "the law") requires VA to furnish hospital care and medical services to Camp Lejeune Veterans and family members with the following conditions even if there is insufficient medical evidence to conclude that such illnesses or conditions are attributable to residence at Camp Lejeune.

Table 1: Fifteen Covered Clinical Conditions

1. Bladder cancer
2. Breast cancer
3. Esophageal cancer
4. Female infertility
5. Hepatic steatosis
6. Kidney cancer
7. Leukemia
8. Lung cancer
9. Miscarriage
10. Multiple myeloma
11. Myelodysplastic syndromes
12. Non-Hodgkin's lymphoma
13. Neurobehavioral effects
14. Renal toxicity
15. Scleroderma

In order to be eligible for care, the law requires that Camp Lejeune Veterans and family members must have served on activity duty at, or resided at, Camp Lejeune for not less than 30 days during the period January 1, 1957, through December 31, 1987. Camp Lejeune Veterans and family members who meet the 30 day requirement will be eligible for the Camp Lejeune program. Veterans can enroll in the VA (Category 6 priority group) and receive comprehensive health care. As category 6 enrollees, Veterans will be responsible for co-pays for all care other than the 15 conditions in the law. Qualifying family members may be reimbursed for healthcare costs related to the 15 illnesses or conditions listed in the law.

The law does not give authority to cover conditions other than the 15 designated conditions and illnesses and "hospital care and medical services may not be furnished…for an illness or condition of a Camp Lejeune Veterans or family member that is found, in accordance with guidelines issued by the Under Secretary for Health, to have resulted from a cause other than the residence" at Camp Lejeune. Clinically indicated screening without co-pays is part of the comprehensive health benefit for enrolled Veterans. VA will reimburse eligible family members for screening that is clinically indicated or recommended by the U.S Preventive Services Task Force that leads to a diagnosis of a covered condition.

According to the law, VA is the payer of last resort for the 15 covered conditions in family member care after all other third party payers. VA will reimburse hospital care or medical services that were provided to family members on, or after, March 26, 2013, which is the date that funds were first appropriated and available to implement medical care provided to family members.

Clinical input required to administer the Camp Lejeune Program.

Clinical input may be needed in administering the Camp Lejeune Program for those conditions that are difficult to define or have multiple potential causes. In those cases where clinical input is necessary, clinicians will use a three-step process to determine whether a medical illness, injury or condition is eligible for coverage under the Camp Lejeune Program. **First,** the clinician must determine whether or not the applicant has one or more of the 15 covered conditions. **Second**, for those conditions with multiple potential causes the clinician must decide whether the condition is probably due to a cause other than exposure to contaminated water at Camp Lejeune. This guidance and clinical algorithms will assist clinicians in making this determination. **Third**, the clinician may sometimes be called upon to decide whether or not an episode of care or treatment is related to one of the 15 covered conditions. This third determination must be made independently for each clinic visit or treatment delivered.

Decision point # 1 - Does the Camp Lejeune Program participant have one or more of the covered conditions?

To establish if a patient has a condition covered by the law, diagnostic criteria and time of onset will need to be evaluated. Some of the 15 conditions have clear evidence-based clinicopathologic diagnostic criteria while others are less clearly defined or supported by existing medical scientific knowledge. Some of the 15 conditions are well defined but have an onset concurrent with exposure and are unlikely to occur years after exposure to contaminated water. Table 2 divides the 15 conditions according to these general categories:

Table 2: Diagnostic Criteria and Time of Onset

Covered conditions with well-established diagnostic criteria	
• Bladder cancer • Breast cancer • Esophageal cancer • Kidney cancer • Leukemia	• Lung cancer • Multiple myeloma • Myelodysplastic syndromes • Non-Hodgkin's lymphoma • Scleroderma
Covered conditions whose onset occurs at the time of solvent exposure	
• Female infertility or miscarriage • Neurobehavioral effects* • Hepatic steatosis • Renal toxicity*	
Covered conditions with poorly defined diagnostic criteria	
• Neurobehavioral effects* • Renal toxicity*	

* Listed twice

Clinicians will consider the medical history and diagnostic criteria in determining if a Camp Lejeune Veteran or family member has a covered condition.

Covered conditions with well-established diagnostic criteria:
Diagnoses such as cancers, myelodysplastic syndrome, and scleroderma are made according to established clinicopathologic diagnostic criteria. Veterans or family members with these diagnoses will be covered by the Camp Lejeune program.

Covered conditions whose onset occurs at the time of solvent exposure:
Hepatic steatosis, female infertility, miscarriage, and neurobehavioral effects have been described in the scientific literature to occur as a result of acute exposure to solvents. The conditions occur <u>during</u> acute exposure or shortly thereafter. Hepatic steatosis and neurobehavioral effects generally resolve after cessation of exposure. It is unlikely that these conditions would have an onset many months or years after exposure to contaminated water at Camp Lejeune. Clinicians will need to review the clinical history or provider report form of the Veteran or family member, respectively, for these conditions and determine the onset of the condition. If a clinician comes to the conclusion that the timing of onset of the covered condition is not consistent with exposure to contaminated water at Camp Lejeune, then VA cannot waive co-payments for Veterans or reimburse care for family members. For example, current scientific studies provide some evidence that solvent exposure during, but not before, pregnancy is associated with miscarriage and that there are no persistent effects of solvent exposure on miscarriage or fetal loss after the acute exposure.

Clinical conditions with poorly defined diagnostic criteria:
Neurobehavioral effects and renal toxicity are not discrete diagnostic entities with commonly recognized criteria or clinicopatholigcal findings. Therefore, for the covered

conditions that are not well defined, clinical judgment guided by the applicant's medical history and diagnostic findings will be needed to determine if a Camp Lejeune Veteran or family member has a covered condition and that the manifestations began during or shortly after the time of exposure at Camp Lejeune. In cases where there is reasonable doubt as to the diagnosis or primary cause for the diagnosis, clinicians should resolve in favor of the Camp Lejeune Veteran or family member. Specific clinical guidance for these conditions is provided below.

Decision point #2 – Is there evidence that the condition occurred as a result of a cause other than residence at Camp Lejeune?

Another provision of the law directs that "hospital care and medical services may not be furnished…for an illness or condition of a family member that is found, in accordance with guidelines issued by the Under Secretary for Health, to have resulted from a cause other than the residence" of the Camp Lejeune Program family member.

Examples of this include:

- Current infertility or miscarriage in a woman who resided at Camp Lejeune as a child.
- Chronic renal disease with onset 25 years after residence at Camp Lejeune in a patient with obstructive uropathy.
- Neuropathy with onset 30 years after residence at Camp Lejeune in a patient with diabetes.
- Hepatic steatosis in an obese patient who requests bariatric surgery.

If a clinician comes to the conclusion that the cause for the covered condition is related to other etiology (cause(s)) not consistent with exposure to contaminated water at Camp Lejeune, then VA cannot waive co-payments for Veterans or reimburse care for family members.

Decision point # 3 - Is the episode of care or treatment related to the covered condition?

Clinicians providing care to Veterans will be asked to verify at each clinical visit if treatments provided to Veterans are related to one of the fifteen Camp Lejeune conditions. This will ensure that Veterans are not charged copays for Camp Lejeune related conditions.

Bills for family member care will be received by the Financial Services Center. In most cases, the Financial Services Center claims processing system should be able to determine if a billed episode of care is related to one of the 15 covered conditions. However in some cases, it will be very difficult to determine if a bill is related to the covered condition. In those cases clinical input will be required.

The following principles will be used by clinicians for decision point # 3.

- Comprehensive coverage during active cancer treatment: According to the medical literature, surgery, chemotherapy and/or radiation therapy for cancer can result in secondary systemic effects on virtually all other organ systems in the body. In recognition of these systemic whole body effects of treatment of the covered cancers, VA will provide family member reimbursement of most medical treatments (as the last payer) during the active treatment phase of these cancers. Similarly, VA will not require Veteran copays for these cancers. VA will not pay for treatments specifically excluded in 38 CFR 17.38 (for example, abortions). The treating oncologist can certify the duration of active treatment or this type of coverage will be provided in six month increments following the initial cancer diagnosis. Once the active treatment of the cancer is completed, VA will provide healthcare reimbursement for treatments only of the covered condition.

- In those situations where the bill for family member care is not itemized by diagnosis, such as a general medical visit to a primary care provider, VA will reimburse for care as a last payer for bills in which the ICD9/10 code of the primary diagnosis is a covered condition.

- Reimbursement of family member bills medical conditions determined to be secondary to a covered condition: VA has determined that it has the legal authority to provide last payer coverage for health care of conditions that are directly caused by one of the 15 covered conditions or their treatment. In order for Camp Lejeune program participants to be reimbursed for health care of a secondary condition, they will need to have their health care providers certify that the secondary condition is caused or exacerbated by the covered condition or its treatment. VA will review the medical evidence provided and make a determination on eligibility for coverage.

GUIDANCE ON COVERED CONDITIONS

As listed in Table 1, Public Law 112-154 requires VA to furnish hospital care and medical services to Camp Lejeune Veterans and family members with 15 covered medical conditions. Clinicians should consider this guidance in conjunction with standard diagnostic criteria to determine whether a Veteran or family member seeks care for one of these conditions. The following section discusses diagnosis of the 15 conditions and related determinations.

Cancer and neoplastic diagnoses:

The law covers eight neoplasms and myelodysplastic syndrome.

- Esophageal cancer
- Lung cancer
- Breast cancer
- Bladder cancer
- Kidney cancer
- Leukemia
- Multiple myeloma
- Myleodysplastic syndromes

- Non-Hodgkins lymphoma

Current scientific research demonstrates that these conditions can be associated with a toxic exposure, can have a long latency period, and can be diagnosed years after the exposure occurred. These conditions have specific clinical and pathological criteria which makes establishing these diagnoses straightforward.

Scleroderma:

Scleroderma has well established diagnostic criteria. The American College of Rheumatology has defined criteria, which are 97% sensitive and 98% specific for systemic sclerosis (SSc) as follows:

Major criterion:

- Proximal diffuse (truncal) sclerosis (skin tightness, thickening, non-pitting induration)

Minor criteria:

- Sclerodactyly (only fingers and/or toes)
- Digital pitting scars or loss of substance of the digital finger pads (pulp loss)
- Bilateral basilar pulmonary fibrosis

The patient should fulfill the major criterion or two of the three minor criteria. Raynaud's phenomenon is observed in 90-98 % of Systemic Scleroderma patients.

Scleroderma can occur at any time following a toxic exposure. Camp Lejeune Veterans and family members who present with scleroderma will be covered by the law.

Miscarriage:

Miscarriage is a layperson's term for a spontaneous abortion. A spontaneous abortion is the naturally occurring expulsion of an embryo or fetus before viability. In clinical practice, spontaneous expulsion of a fetus that weighs less than 500 grams or is at or before 20 weeks gestation (or 18 weeks after fertilization) is considered a spontaneous abortion or miscarriage. Spontaneous expulsion of a fetus that is beyond 20 weeks gestation or weighs more than 500 grams is considered a preterm birth. (Sources: American Society for Reproductive Medicine, Stedman's medical dictionary, Medline Plus, HHS Office of Women's Health)

Spontaneous abortion is the most common complication of early pregnancy. About 15% to 20% of known pregnancies end in spontaneous abortion. With the use of serial human chorionic gonadotropin (hCG) measurements to detect early subclinical pregnancy losses, the percentage increases to 30%. About 80% of spontaneous pregnancy losses occur in the first trimester; the incidence decreases with each gestational week.

Most spontaneous abortions are caused by chromosome abnormalities that occur spontaneously and are not related to the mother's or father's genetic make-up. Other factors that may increase the risk for spontaneous abortion include: drug and alcohol abuse, smoking, exposure to environmental toxins, hormone level abnormalities, infection, obesity, physical abnormalities of the uterus or cervix, and some chronic systemic diseases (like diabetes and other autoimmune diseases).

Current scientific studies provide some evidence that solvent exposure during, but not before, pregnancy is associated with miscarriage. Therefore miscarriages that occurring during the period 1957-87 while the Veteran or family member lived on Camp Lejeune shall be covered if they require ongoing medical treatment. Current scientific evidence suggests that there are no persistent effects of solvent exposure on miscarriage or fetal loss (National Research Council (NRC) *Contaminated Water Supplies at Camp Lejeune: Assessing Potential Health Effects*. p. 182). Clinicians should carefully assess whether continued health care is needed for chronic, persistent medical problems associated with a miscarriage that occurred during solvent exposure at Camp Lejeune.

Female Infertility:

The current scientific literature suggests an association between concurrent exposure to solvents and reduced fecundity (a woman's ability to become pregnant). Infertility and any associated chronic persistent medical problems that occurred during the period 1957 to 1987 while the Veteran or family member lived on Camp Lejeune are covered. There is currently no scientific evidence to support an association with chronic female infertility after cessation of exposure to solvents (NRC p. 181). Similarly the NRC report found no evidence that exposure to organic solvents while in-utero increases the risk for adverse fertility effects as a reproductively mature adult.

Neurobehavioral effects:

According to the current scientific literature, neurobehavioral symptoms associated with solvent exposure have included acute decrements in concentrating ability and visuospatial skills, fine motor abnormalities, at, or shortly after, solvent exposure. Neurobehavioral symptoms secondary to the contaminant levels in the water at Camp Lejeune would likely have been manifest at the time of exposure or shortly thereafter.

In assessing whether current neurobehavioral effects that require medical treatment are associated with exposure to water contaminants at Camp Lejeune, clinicians should evaluate the onset and duration of symptoms to determine if the Veteran or family member has symptoms that are as likely as not, related to exposure to volatile organic compounds in the past. The current evidence suggests that neurobehavioral effects occurring after a long asymptomatic period are not likely to be secondary to the contaminated water at Camp Lejeune. The scientific literature to date indicates that with the type of exposure at Camp Lejeune (contaminated water used by adults for activities such as drinking and bathing), and its remoteness in time (the exposure occurred more decades ago); it is unlikely that neurobehavioral effects would have persisted and would require treatment at this time.

Organic solvent exposure has been associated with various forms of central nervous system toxicity. There is a widespread agreement that chronic effects are seen primarily after long-term, high-level occupational exposure, with objective testing showing decrements in concentrating ability, visuospatial skills, and fine motor abnormalities, after 10 years or more of occupational exposure (Flodin 1984). Such long-term exposures are also associated with the development of personality changes (Chen 2001). Earlier, mild disease may be seen after as little as three years of exposure. Some exposed individuals develop peripheral neuropathy with acute onset at the time of exposure that can persist. Mild disease, with acute symptoms, generally resolves after cessation of exposure and has not been associated with progressive disease (van Valen 2009). Low-level exposures are not associated with adverse long-term cognitive outcomes (Dick 2010). Research to date has not shown any evidence of progression or worsening after cessation of exposure. There are also no known cases of onset of symptoms after cessation of exposure.

Current classification systems to define neurobehavioral effects are summarized below:

Type 1: Acute symptoms only, including impairment of memory, poor concentration, fatigue, and decreased motivation. In general these resolve rapidly after cessation of exposure. Headaches are not commonly included in a listing of symptoms, because nonspecific headaches are known to occur frequently in the general population.
Type 2A: A sustained change in mood and / or personality, with reduced motivation, poor impulse control, irritability and often anxiety is seen after longer term exposures. This form is generally not associated with performance decrements
Type 2B: Impairment in intellectual function that is associated with cognitive deficits, including problems with attention, concentration, visuospatial skills, and verbal memory. In addition, fine motor performance can be impaired.

Type 3: Severe chronic toxic encephalopathy is characterized by global deterioration in cognitive functions and memory.

Type 1 and 2 disorders are the most likely to be reported among solvent-exposed workers. Type 3 disorders to date have been seen only in individuals who have abused solvent-containing products (i.e., by deliberately inhaling organic solvent vapors for their euphoric properties (NIOSH 1987).

The scientific literature demonstrates that there is inadequate information to associate exposure to solvent contaminated water with Alzheimer's disease, amyotrophic lateral sclerosis, attention deficit/hyperactivity disorder, multiple sclerosis, reductions in color discrimination, hearing and olfactory functions. While these conditions may have symptoms that overlap with those of neurobehavioral effects, the law does not include coverage for other diagnosable neurologic diseases.

Renal toxicity:

In evaluating whether a Camp Lejeune Veteran or family member has renal toxicity due to solvent exposure, the clinician should perform a thorough assessment of the patient's history and onset of chronic kidney disease and other comorbid medical conditions. Once a diagnosis is made, the clinician should consider whether it is probable that the patient's kidney disease resulted from a known etiology other than solvent toxicity. Chronic kidney disease has many known etiologies. The most common causes of chronic kidney disease include long term diabetes and hypertension. Conditions that might be excluded as a covered condition due to renal toxicity from a Camp Lejeune exposure would include a patient with forms of chronic kidney disease with another known etiology such as those with underlying diabetic nephropathy, obstructive uropathy, hypertensive nephrosclerosis, sickle cell kidney disease, HIV-associated nephropathy, drug-induced kidney disease, etc. Reviewing clinicians should determine if a patient's clinical course of chronic kidney disease is associated with a known cause, e.g. a 20 year history of diabetes. A clinician might reasonably conclude that the renal disease is as likely as not associated with the patient's diabetes and they would not be covered by the law. If a patient's clinical course of kidney disease appears atypical, in that their progression of kidney disease is faster than expected, then exacerbation by TCE, PCE or other organic solvents in the contaminated water should be considered.

Hepatic steatosis:

Hepatic steatosis is not a disease rather it is a common pathological finding in medical conditions that affect the liver. In western countries, it affects up to one third of the population and up to 75% in some subgroups such as obese patients. Hepatic steatosis or simple fatty liver can be caused by a variety of conditions including alcohol abuse, overweight or obesity, Type 2 Diabetes, metabolic syndrome, medication use and hepatitis.

Generally hepatic steatosis or fatty liver resolves by treating the underlying condition. In evaluating whether a Camp Lejeune Veteran or family member has hepatic steatosis related to Camp Lejeune, the clinician should consider whether it is more likely than not that the patient has fatty liver disease from a known etiology. The most common causes of hepatic steatosis include obesity and alcohol abuse. If a patient's clinical course is consistent with a known cause of hepatic steatosis, their treatment for hepatic steatosis should not be covered by the law.

Camp Lejeune Veterans and family members with hepatic steatosis of unclear or unknown etiology should be covered by the Camp Lejeune law. Moreover, if a patient's clinical course of hepatic steatosis is atypical or progresses faster than expected, then exacerbation by TCE, PCE or other organic solvents in the contaminated water should be considered.

Points of contact for this guidance:

The VA Camp Lejeune Task Force is co-led by the VHA Office of Public Health and the VHA Chief Business Office (CBO), with representation from the VA Office of General Counsel, VA Office of Congressional and Legislative Affairs, VA Office of Public and Intergovernmental Affairs, VHA Office of Primary Care in the Office of Patient Care Services, VHA Communications, and other offices. The main VA Web site for Camp Lejeune historic water contamination and potential health concerns is http://www.publichealth.va.gov/exposures/camp-Lejeune; individuals can subscribe to email updates. The Marines' registry appears on their Camp Lejeune Historic Drinking Water site: https://clnr.hqi.usmc.mil/clwater/index.html. The ATSDR site is http://www.atsdr.cdc.gov/sites/lejeune/ The Task Force co-chairs are Terry Walters, MD, MPH, Office of Public Health, 202-461-1020; Katie Shebesh, Chief Business Office, 202-461-1600.

Selected Reading:

General

Agency for Toxic Substances and Disease Registry, U.S. Department of Health and Human Services, Atlanta GA. Available:http://www.atsdr.cdc.gov/sites/lejeune/ [accessed September 4, 2013].

Committee on Contaminated Water at Camp Lejeune; National Research Council. *Contaminated Water Supplies at Camp Lejeune: Assessing Potential Health Effects*. 2009. Washington, DC: The National Academies Press. Available: http://www.nap.edu/catalog.php?record_id=12618

Committee on Human Health Risks of Trichloroethylene; National Research Council. *Assessing the Human Health Risks of Trichloroethylene: Key Scientific Issues.* Washington, DC: The National Academies Press , 2006. Available: http://www.nap.edu/catalog/11707.html

Committee on Gulf War and Health. Institute of Medicine. 2003. Gulf War and Health: Volume 2. Insecticides and Solvents:: Literature Review of Pesticides and Solvents. Washington, DC: The National Academies Press, 2003. Available: http://www.nap.edu/catalog/10628.html .

Honoring America's Veterans and Caring for Camp Lejeune Families Act of 2012. Available:http://www.gpo.gov/fdsys/pkg/BILLS-112hr1627enr/pdf/BILLS-112hr1627enr.pdf. [accessed September 4, 2013]

Hepatic Steatosis:

Mishra P, Younossi ZM. **2007.** Current treatment strategies for non-alcoholic fatty liver disease (NAFLD). Curr Drug Discov Technol. 4(2):133-40.

Raman M, Allard J. 2006. Nonalcoholic fatty liver disease: A clinical approach and review. Can J Gastroenterol. 20(5): 345–349.

Younossi ZM. 2008. Review article: current management of non-alcoholic fatty liver disease and non-alcoholic steatohepatitis. Aliment Pharmacol Ther.28(1):2-12.

Neurobehavioral effects:

Attarchi MS, Labbafinejad Y, Mohammadi S. 2010. Occupational exposure to different levels of mixed organic solvents and colour vision impairment. Neurotoxicol Teratol. 32(5):558-62.

Baker and Seppalainen 1986. International Solvent Workshop, Raleigh, N.C. October 1985. ; Anonymous. 1986. Human aspects of solvent neurobehavioral effects (Raleigh Workshop). Neurotoxicology 7:43-56.

Chen R, Dick F, Semple S, Seaton A, Walker LG. 2001. Exposure to organic solvents and personality. Occup Environ Med. 58(1):14-8.

Costa TL, Barboni MT, Moura AL, Bonci DM, Gualtieri M, de Lima Silveira LC, Ventura DF. 2012. Long-term occupational exposure to organic solvents affects color vision, contrast sensitivity and visual fields. PLoS One. 7(8):e42961

Dick FD, Bourne VJ, Semple SE, Fox HC, Miller BG, Deary IJ, Whalley LJ. 2010. Solvent exposure and cognitive ability at age 67: a follow-up study of the 1947 Scottish Mental Survey. Occup Environ Med. 67(6):401-7.

Flodin U, Edling C, Axelson O. 1984. Clinical studies of psychoorganic syndromes among workers with exposure to solvents. Am J Ind Med.5:287-95.

Janulewicz PA, White RF, Winter JM, et al. 2008. Risk of learning and behavioral disorders following prenatal and early postnatal exposure to tetrachloroethylene (PCE)-contaminated drinking water. Neurotoxicol. Teratol. 30(3):175-185.

Kilburn KH, Warshaw RH. 1993. Effects on neurobehavioral performance of chronic exposure to chemically contaminated well water. Toxicol Ind Health. 9(3):391-404.

Meyer-Baron M, Blaszkewicz M, Henke H, Knapp G, Muttray A, Schäper M, van Thriel C. 2008. The impact of solvent mixtures on neurobehavioral performance: conclusions from epidemiological data. Neurotoxicology29:349-60.

NIOSH. Organic Solvent Neurotoxicity. Current Intelligence Bulletin 48. DHHS (NIOSH) Publication Number 87-104 (http://www.cdc.gov/niosh/docs/87-104/)

Reif JS, Burch JB, Nuckols JR, Metzger L, Ellington D, Anger WK. 2003. Neurobehavioral effects of exposure to trichloroethylene through a municipal water supply. Environ Res. 93(3):248-58.

van Valen E, Wekking E, van der Laan G, Sprangers M, van Dijk F. 2009. The course of chronic solvent induced encephalopathy: a systematic review. Neurotoxicology. 30:1172-86.

Zamyslowska-Szmytke E, Politanski P, Sliwinska-Kowalska M. 2011. Balance system assessment in workers exposed to organic solvent mixture. J Occup Environ Med. 53(4):441-7.

Renal Toxicity

Brosnahan G, Fraer M. 2010. Chronic kidney disease: whom to screen and how to treat, part 1: definition, epidemiology, and laboratory testing. South Med J. 103(2):140-6.

Calvert GM, Ruder AM, Petersen MR. 2011. Mortality and end-stage renal disease incidence among dry cleaning workers. Occup Environ Med.68(10):709-16.

Jacob S, Héry M, Protois JC, Rossert J, Stengel B. 2007. Effect of organic solvent exposure on chronic kidney disease progression: the GN-PROGRESS cohort study. J Am Soc Nephrol. Jan;18(1):274-81.

NIDDK. National Institutes of Health. U.S. Department of Health and Human Services. Kidney Disease Statistics for the US. Available: http://kidney.niddk.nih.gov/kudiseases/pubs/kustats/KU_Diseases_Stats_508.pdf. [Downloaded 1 March 2013]

Scleroderma:

LeRoy EC, Medsger TA, Jr. **2001.** Criteria for the classification of early systemic sclerosis. J Rheumatol. 28(7):1573-6.

Mora GF**. 2009.** Systemic sclerosis: environmental factors. J Rheumatol. 36(11):2383-96.

Kettaneh A, Al Moufti O, Tiev KP, Chayet C, Toledano C, Fabre B, et al. **2007.** Occupational exposure to solvents and gender-related risk of systemic sclerosis: a metaanalysis of case-control studies. J Rheumatol.34(1):97-103.

Guidance for VHA Staff Honoring America's Veterans and Caring for Camp Lejeune Families Act of 2012, Covered Clinical Conditions

CLINICAL ALGORITHM

The clinical algorithm incorporates the information presented in the guidance in a format that maximally facilitates clinical decision-making. The algorithm can be used as a structured approach in assessing the clinical eligibility of an applicant to the Camp Lejeune program.

The **Core Algorithm** is a pictorial description of the key steps in evaluating if the applicant has one or more of the conditions covered by the law. The Core Algorithm refers to four sub-algorithms used to identify if there is evidence that the condition occurred as a result of a caus other than exposure to contaminated water during residence at Camp Lejeune.

Standardized symbols are used to display each step in the algorithm. Arrows connect the numbered boxes indicating the order in which the steps should be followed. Rounded rectangles represent a clinical state or condition. Hexagons represent a decision point, formulated as a question that can be answered "Yes" or "No". A horizontal arrow points to the next step if the answer is YES. A vertical arrow should be followed for a negative answer. Rectangles represent an action in the assessment process. Ovals represent a link forward to another section (i.e. sub-algorithm) or a link back to the Core Algorithm. A letter within a box (e.g., **C- 1**) refers to text annotation following the flowchart.

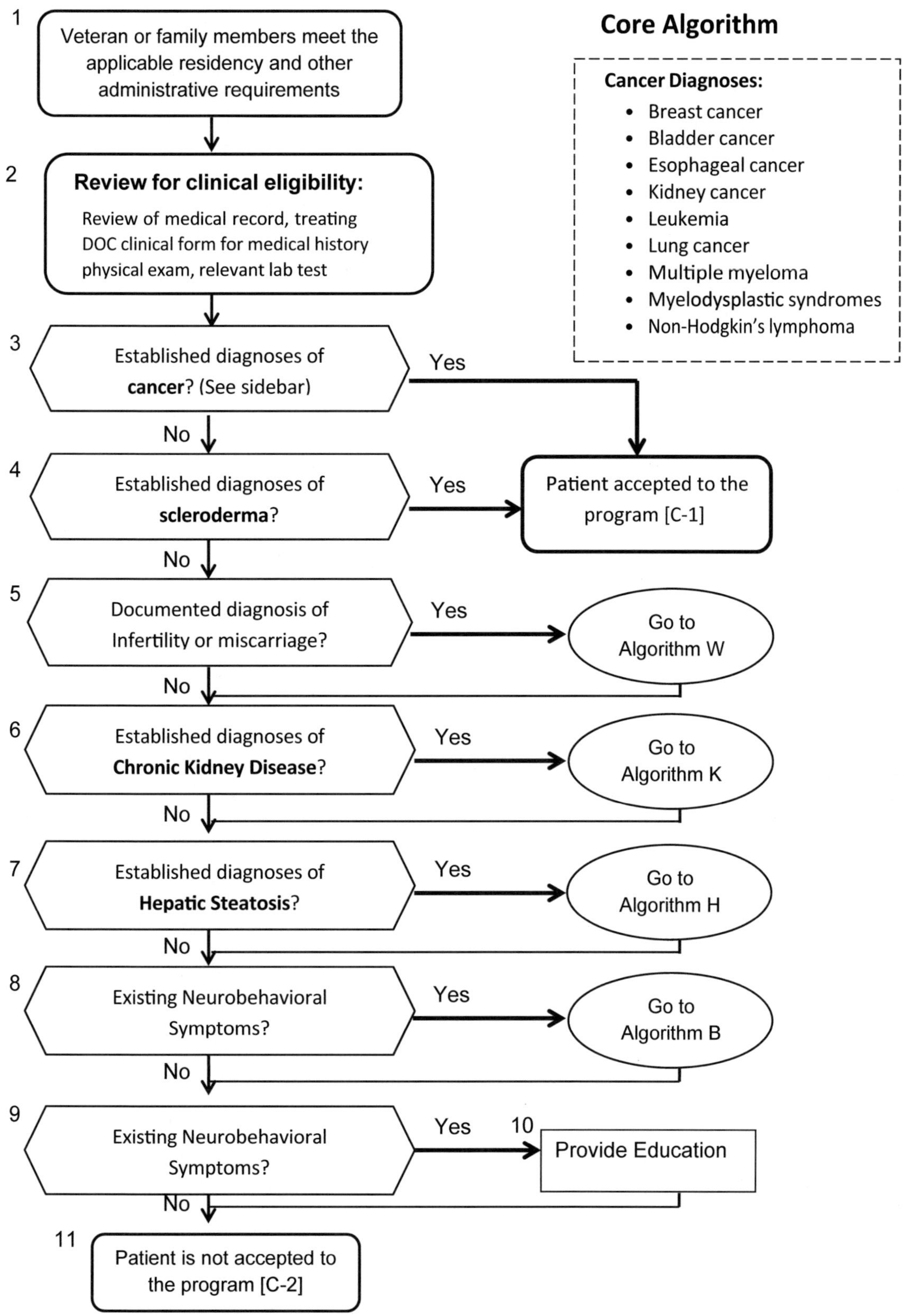
Core Algorithm
1
Veteran or family members meet the applicable residency and other administrative requirements
2
Review for clinical eligibility:
Review of medical record, treating DOC clinical form for medical history physical exam, relevant lab test
3
Established diagnoses of cancer? (See sidebar)
Yes
No
Cancer Diagnoses:
• Breast cancer
• Bladder cancer
• Esophageal cancer
• Kidney cancer
• Leukemia
• Lung cancer
• Multiple myeloma
• Myelodysplastic syndromes
• Non-Hodgkin's lymphoma
4
Established diagnoses of scleroderma?
Yes
Patient accepted to the program [C-1]
No
5
Documented diagnosis of Infertility or miscarriage?
Yes
Go to Algorithm W
No
6
Established diagnoses of Chronic Kidney Disease?
Yes
Go to Algorithm K
No
7
Established diagnoses of Hepatic Steatosis?
Yes
Go to Algorithm H
No
8
Existing Neurobehavioral Symptoms?
Yes
Go to Algorithm B
No
9
Existing Neurobehavioral Symptoms?
Yes
10
Provide Education
No
11
Patient is not accepted to the program [C-2]

Annotations --- Core Algorithm

C1 --- Applicant has a confirmed diagnosis of cancer or scleroderma. Applicant is clinically eligible for the Camp Lejeune program.

C2 --- Applicant is administratively eligible for the Camp Lejeune program (has resided in camp Lejeune for at least 30 days between Jan 1, 1957 and Dec 31, 1987, but does not yet have any of the 15 medical conditions specified in the law.

Camp Lejeune Program
Kidney Toxicity (CKD)
Algorithm

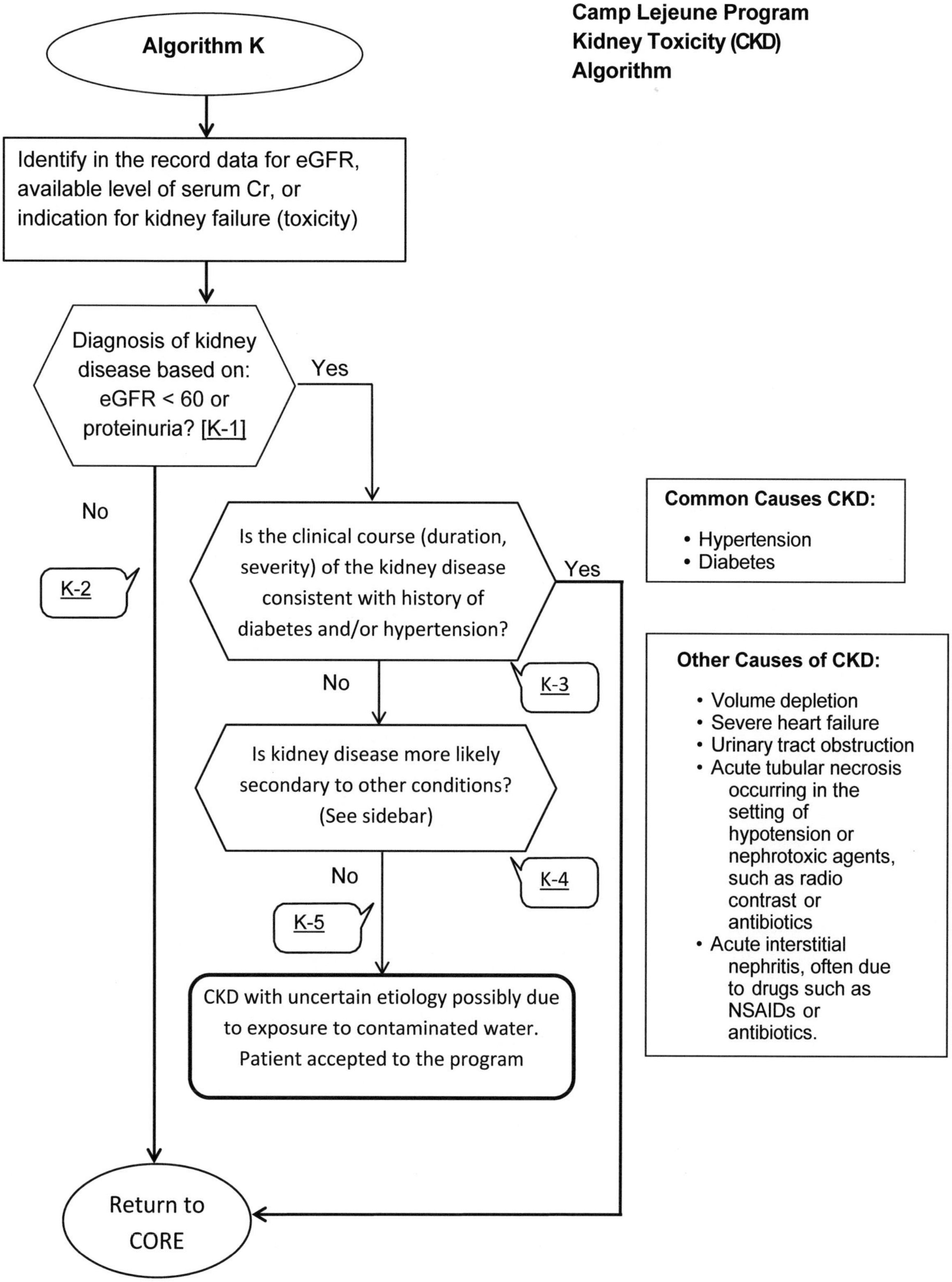

Annotations --- Kidney Toxicity

K1 -- Diagnosis of kidney disease: Applicant has a history of renal toxicity or kidney disease concurrent with Camp Lejeune residence or shortly after the time of possible exposure to contaminated water at Camp Lejeune.

The two most common causes of chronic kidney disease (CKD) are diabetes and hypertension. In most instances, it will be possible to identify the most likely cause of CKD using history, physical examination, laboratory testing and imaging tests. A kidney biopsy should be considered for patients with nephrotic range proteinuria (urine to creatinine ratio >3.5), particularly in the absence of diabetes, to determine the histopathology of the kidney disease.

K2 -- Applicant is still administratively eligible for the Camp Lejeune Program but does not have evidence of renal toxicity as a covered condition.

K3 -- Applicant has no history of renal toxicity or kidney disease concurrent with Camp Lejeune residence or around the time of possible exposure to contaminated water at Camp Lejeune. Applicant has evidence of kidney disease due to long standing diabetes or refractory hypertension, which are common causes of kidney failure and are not related to exposure to the contaminants in the water at Camp Lejeune. Current kidney disease is due to another cause other than exposure to contaminated water at Camp Lejeune. Applicant is does not have a covered condition eligible for coverage by the Camp Lejeune Program at this time.

In most patients with diabetes, CKD should be attributable to diabetes if:

- Macroalbuminuria is present; or
- Microalbuminuria is present
    - in the presence of diabetic retinopathy,
    - in type 1 diabetes of at least 10 years' duration

Other cause(s) of CKD should be considered in the presence of any of the following circumstances:

- Absence of diabetic retinopathy;
- Low or rapidly decreasing GFR;
- Rapidly increasing proteinuria or nephrotic syndrome;
- Presence of active urinary sediment;
- Signs or symptoms of other systemic disease; or
- >30% reduction in GFR within 2-3 months after initiation of an ACE inhibitor or ARB.

K4-- Applicant has no history of renal toxicity or kidney disease concurrent with Camp Lejeune residence or around the time of possible exposure to contaminated water at Camp Lejeune. Applicant has evidence of kidney disease consistent with a secondary condition that is not related to exposure to the contaminants in the water at Camp Lejeune. Current kidney disease is due to another cause other than exposure to contaminated water at Camp Lejeune. Applicant does not have a covered condition eligible for coverage by the Camp Lejeune Program at this time.

K5-- Applicant has Chronic Kidney Disease of uncertain etiology, possibly related to exposure to contaminated water at Camp Lejeune. Applicant has kidney disease of uncertain etiology possibly related to exposure to contaminated water at Camp Lejeune. Applicant has a covered condition, renal toxicity, and is accepted into the Camp Lejeune Program.

Camp Lejeune Program
Hepatic Steatosis Algorithm

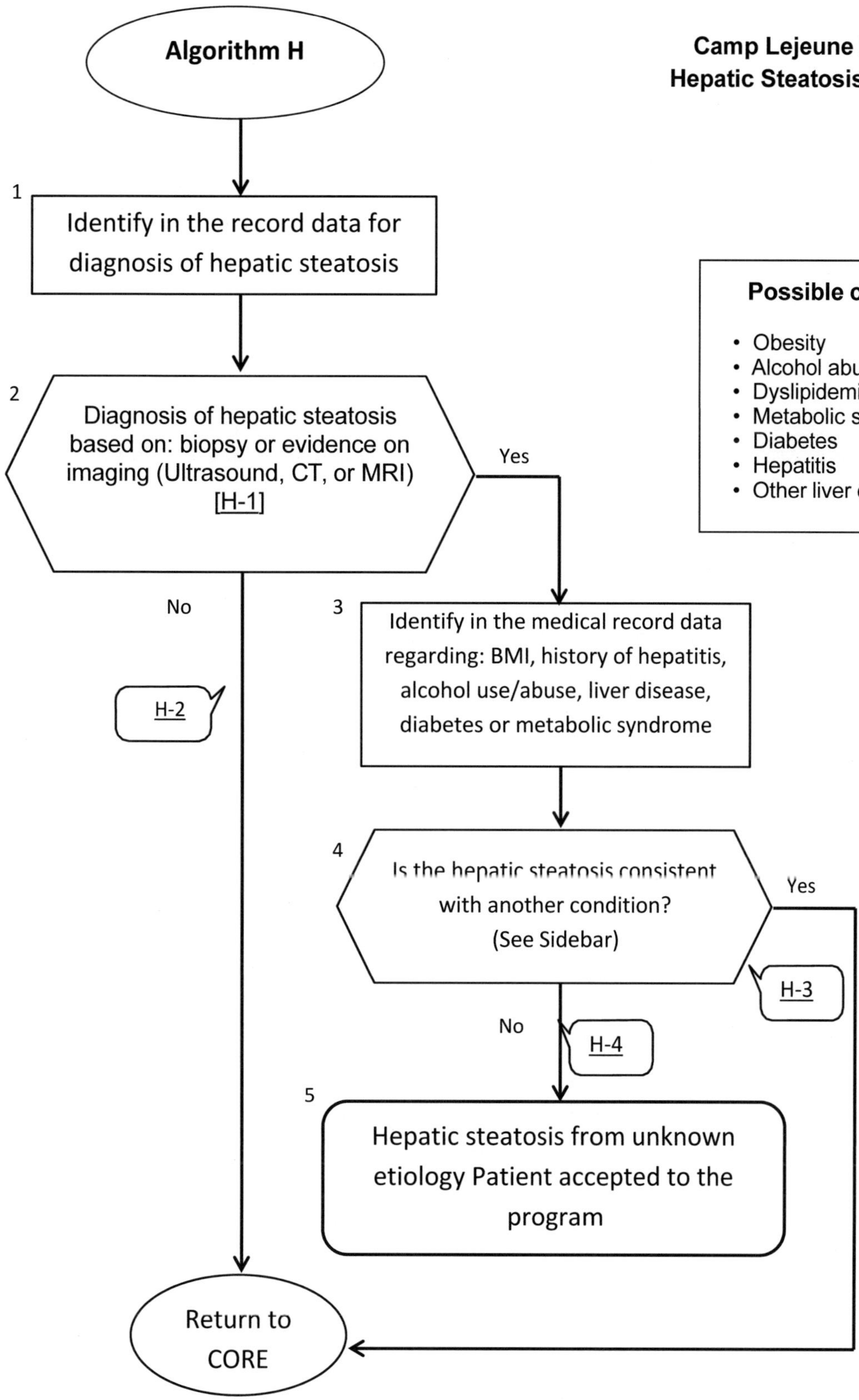

Possible causes:

- Obesity
- Alcohol abuse
- Dyslipidemia
- Metabolic syndrome
- Diabetes
- Hepatitis
- Other liver diseases

Annotations --- Hepatic Steatosis

H1 – Hepatic Steatosis (fatty liver) is an accumulation of lipids (triglycerides and other lipids) in the liver hepatocytes. Patients are often asymptomatic. It is often diagnosed as an incidental finding on routine medical exams with blood tests revealing abnormal liver function tests. The most common known causes of steatosis are obesity and alcohol abuse. Other possible causes include dyslipidemia, metabolic syndrome, diabetes, hepatitis or other liver disease. Liver biopsy is the only definitive test to confirm diagnosis, exclude other causes, assess extent and predict prognosis. In most instances, it will be possible to identify the existence of hepatic steatosis and define the extent of the condition using diagnostic imaging techniques (Ultrasound, CT or MRI).

H2 -- Applicant does not have clinical evidence (positive biopsy, CT, MRI, or ultrasound test) of hepatic steatosis at this time.

H3 -- Applicant has clinical evidence of hepatic steatosis due to a cause other than exposure at Camp Lejeune. Fatty liver disease can be divided into two main categories: alcohol-related fatty liver disease and non-alcoholic fatty liver disease (NAFLD). A threshold of <20 gm alcohol per day in women and <30 gm in men suggests a diagnosis of NAFLD. NAFLD is associated with obesity, abnormal glucose tolerance and dyslipidemia, and has been described as the hepatic manifestation of the metabolic syndrome. Fatty liver develops in 46-90% of heavy alcohol users, and in up to 94% of obese individuals.

Current hepatic steatosis is due to another cause other than exposure to contaminated water at Camp Lejeune. Applicant is does not have a covered condition eligible for coverage by the Camp Lejeune Program at this time.

H4 -- Applicant has Hepatic Steatosis of unknown etiology. Applicant is accepted into the Camp Lejeune program.

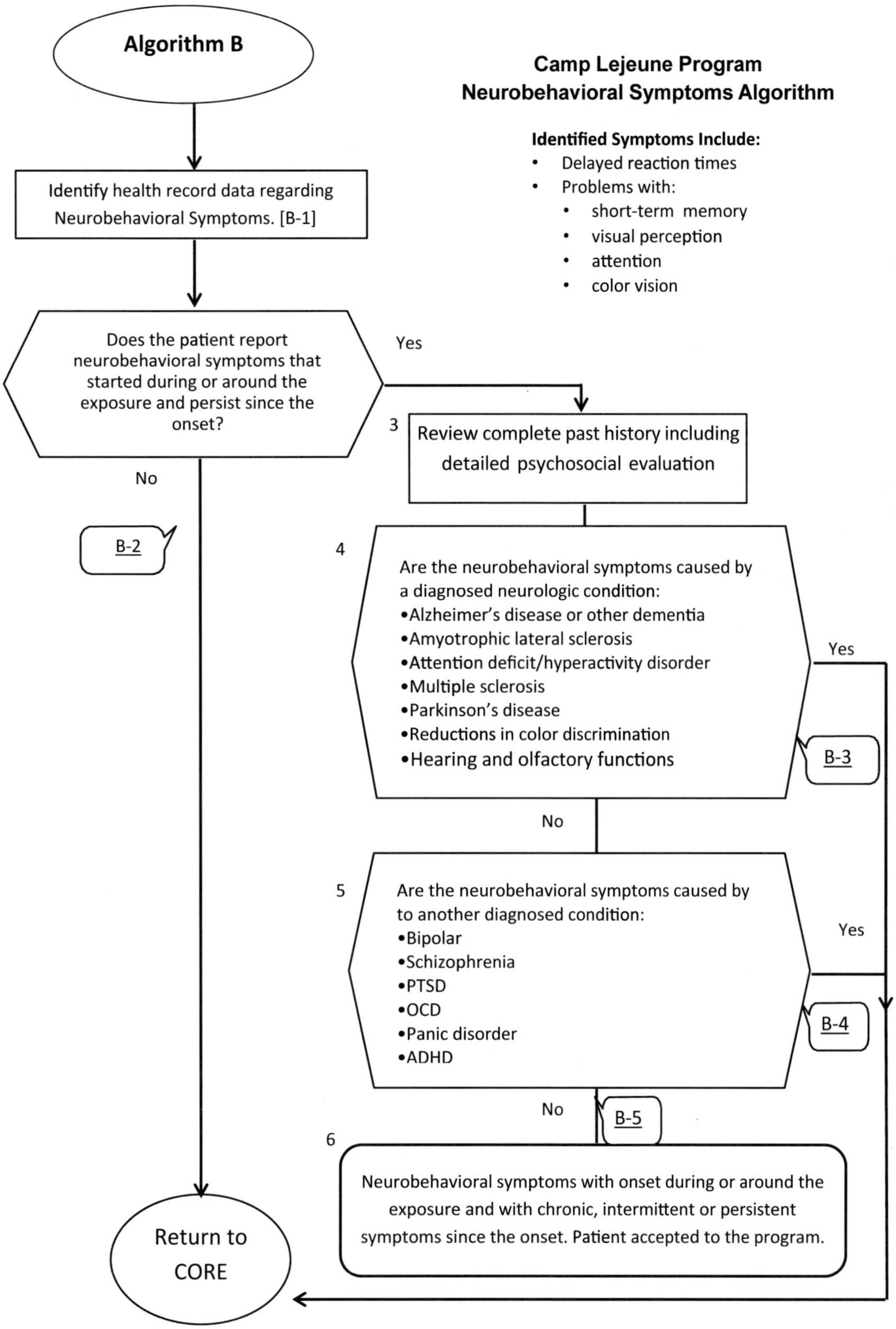
Algorithm B
Camp Lejeune Program
Neurobehavioral Symptoms Algorithm
Identified Symptoms Include:
• Delayed reaction times
• Problems with:
• short-term memory
• visual perception
• attention
• color vision
1
Identify health record data regarding Neurobehavioral Symptoms. [B-1]
2
Does the patient report neurobehavioral symptoms that started during or around the exposure and persist since the onset?
Yes
No
B-2
3
Review complete past history including detailed psychosocial evaluation
4
Are the neurobehavioral symptoms caused by a diagnosed neurologic condition:
•Alzheimer's disease or other dementia
•Amyotrophic lateral sclerosis
•Attention deficit/hyperactivity disorder
•Multiple sclerosis
•Parkinson's disease
•Reductions in color discrimination
•Hearing and olfactory functions
Yes
B-3
No
5
Are the neurobehavioral symptoms caused by to another diagnosed condition:
•Bipolar
•Schizophrenia
•PTSD
•OCD
•Panic disorder
•ADHD
Yes
B-4
No
B-5
6
Neurobehavioral symptoms with onset during or around the exposure and with chronic, intermittent or persistent symptoms since the onset. Patient accepted to the program.
Return to CORE

Annotations --- Neurobehavioral Symptoms

B1 – Identified neurobehavioral symptoms include delayed reaction times, short-term memory impairment, visual perception problems, decreased attention and problems with color vision.

B2 --Applicant did not have symptoms at the time of exposure or documented symptoms first occurred a prolonged time after residence at Camp Lejeune ceased. Research to date has not shown any evidence of onset or progression of symptoms after cessation of exposure.

Applicant does not have neurobehavioral symptoms as a covered condition and is not eligible for the Camp Lejeune Program at this time.

B3 --- Applicant has a neurological diagnosis with neurobehavioral symptoms that are commonly caused by this diagnosis. IOM reviews of solvent exposures in 2003 and 2008 have found inadequate or insufficient evidence of an association between these neurological diagnoses and exposure to the chemicals in the water at Camp Lejeune

B4--- Applicant has a psychiatric diagnosis that causes neurobehavioral symptoms. IOM reviews of solvent exposures in 2003 and 2008 have found inadequate or insufficient evidence of an association between these psychiatric diagnoses and exposure to the chemicals in the water at Camp Lejeune.

B5 --- Applicant has evidence of neurobehavioral symptoms whose onset during or around their exposure at Camp Lejeune. A chronic, intermittent, or persistent symptom since their exposure suggests neurobehavioral effects secondary to exposure at Camp Lejeune. Applicant accepted into the Camp Lejeune program.

Camp Lejeune Program Algorithm Reproductive Health: Miscarriages and Infertility in Women

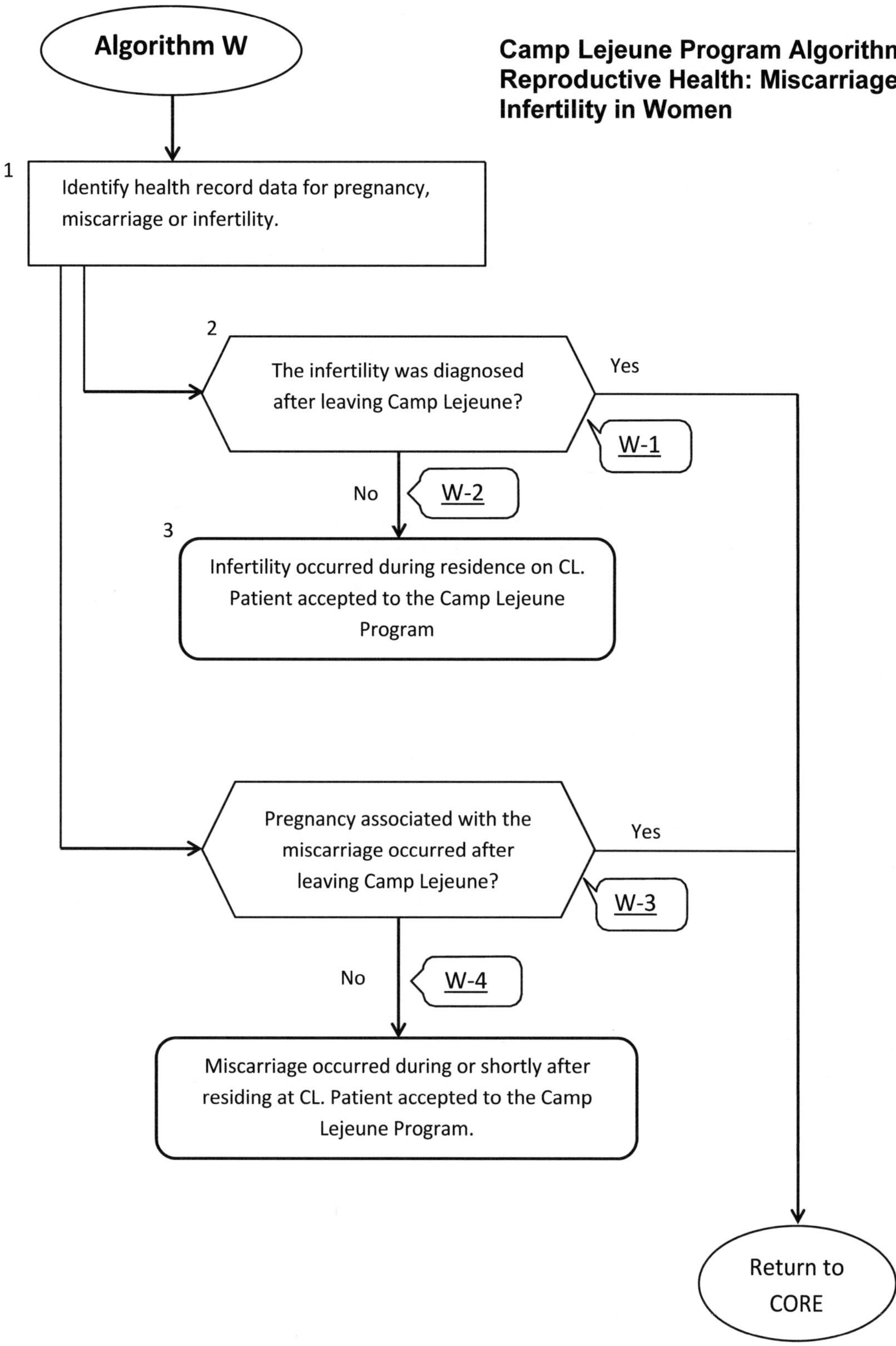

Annotations--- Female Health:

W1 - Infertility was diagnosed after leaving Camp Lejeune. There is currently no scientific evidence to support an association with chronic female infertility after cessation of exposure to solvents. Similarly the NRC report found no evidence that exposure to organic solvents while in-utero increases the risk for adverse fertility effects as a reproductively mature adult. Applicant does not have female infertility that is covered by the Camp Lejeune program.

W2 -Applicant has medical complications requiring continued medical treatment from female infertility while exposed to contaminated water at Camp Lejeune. Medical condition is related to infertility during residence at Camp Lejeune. Applicant accepted into the Camp Lejeune program.

W3 - Miscarriage occurred after leaving Camp Lejeune. Current scientific evidence suggests that there are no persistent effects of solvent exposure on miscarriage or fetal loss. Applicant does not have a miscarriage that is covered by the Camp Lejeune program. Applicant is not accepted into the Camp Lejeune Program at this time.

W4 --Applicant has medical complications requiring continued medical treatment from a miscarriage while exposed to contaminated water at Camp Lejeune. Clinicians should carefully assess whether continued health care is needed for chronic, persistent medical problems associated with a miscarriage that occurred during solvent exposure at Camp Lejeune; if care needs persist applicant is accepted into the Camp Lejeune Program.

Appendix C

Excerpt from the Honoring America's Veterans and Caring for Camp Lejeune Families Act of 2012 (P.L. 112-154)

TITLE I—HEALTH CARE MATTERS

SEC. 101. SHORT TITLE.

This title may be cited as the "Janey Ensminger Act."

SEC. 102. HOSPITAL CARE AND MEDICAL SERVICES FOR VETERANS STATIONED AT CAMP LEJEUNE, NORTH CAROLINA.

(a) Hospital Care and Medical Services for Veterans—

(1) IN GENERAL—Paragraph (1) of section 1710(e) is amended by adding at the end the following new subparagraph:

"(F) Subject to paragraph (2), a veteran who served on active duty in the Armed Forces at Camp Lejeune, North Carolina, for not fewer than 30 days during the period beginning on January 1, 1957, and ending on December 31, 1987, is eligible for hospital care and medical services under subsection (a)(2)(F) for any of the following illnesses or conditions, notwithstanding that there is insufficient medical evidence to conclude that such illnesses or conditions are attributable to such service:

"(i) Esophageal cancer.
"(ii) Lung cancer.
"(iii) Breast cancer.
"(iv) Bladder cancer.
"(v) Kidney cancer.
"(vi) Leukemia.
"(vii) Multiple myeloma.
"(viii) Myleodysplasic syndromes.
"(ix) Renal toxicity.
"(x) Hepatic steatosis.

"(xi) Female infertility.
"(xii) Miscarriage.
"(xiii) Scleroderma.
"(xiv) Neurobehavioral effects.
"(xv) Non-Hodgkin's lymphoma."

(2) LIMITATION—Paragraph (2)(B) of such section is amended by striking "or (E)" and inserting "(E), or (F)."

(b) Family Members—

(1) IN GENERAL—Subchapter VIII of chapter 17 is amended by adding at the end the following new section:
—"Sec. 1787. Health care of family members of veterans stationed at Camp Lejeune, North Carolina

"(a) In General—Subject to subsection (b), a family member of a veteran described in subparagraph (F) of section 1710(e)(1) of this title who resided at Camp Lejeune, North Carolina, for not fewer than 30 days during the period described in such subparagraph or who was in utero during such period while the mother of such family member resided at such location shall be eligible for hospital care and medical services furnished by the Secretary for any of the illnesses or conditions described in such subparagraph, notwithstanding that there is insufficient medical evidence to conclude that such illnesses or conditions are attributable to such residence.

"(b) Limitations—(1) The Secretary may only furnish hospital care and medical services under subsection (a) to the extent and in the amount provided in advance in appropriations Acts for such purpose.

"(2) Hospital care and medical services may not be furnished under subsection (a) for an illness or condition of a family member that is found, in accordance with guidelines issued by the Under Secretary for Health, to have resulted from a cause other than the residence of the family member described in that subsection.

"(3) The Secretary may provide reimbursement for hospital care or medical services provided to a family member under this section only after the family member or the provider of such care or services has exhausted without success all claims and remedies reasonably available to the family member or provider against a third party (as defined in section 1725(f) of this title) for payment of such care or services, including with respect to health—plan contracts (as defined in such section)."

(2) CLERICAL AMENDMENT—The table of sections at the beginning of such chapter is amended by inserting after the item relating to section 1786 the following new item:

"1787. Health care of family members of veterans stationed at Camp Lejeune, North Carolina."

(c) Annual Reports—

(1) IN GENERAL—Not later than December 31 of each of 2013, 2014, and 2015, the Secretary of Veterans Affairs shall submit to the Committee on Veterans" Affairs of the Senate and the Committee on Veterans" Affairs of the House of Representatives a report on the care and services provided under sections 1710(e)(1)(F) and 1787 of title 38, United States Code (as added by subsections (a) and (b)(1), respectively).

(2) ELEMENTS—Each report under paragraph (1) shall set forth the following:

(A) The number of veterans and family members provided hospital care and medical services under the provisions of law specified in paragraph (1) during the period beginning on October 1, 2012, and ending on the date of such report.

(B) The illnesses, conditions, and disabilities for which care and services have been provided such veterans and family members under such provisions of law during that period.

(C) The number of veterans and family members who applied for care and services under such provisions of law during that period but were denied, including information on the reasons for such denials.
(D) The number of veterans and family members who applied for care and services under such provisions of law and are awaiting a decision from the Secretary on eligibility for such care and services as of the date of such report.

(d) Effective Date—

(1) IN GENERAL—The provisions of this section and the amendments made by this section shall take effect on the date of the enactment of this Act.

(2) APPLICABILITY—Subparagraph (F) of section 1710(e)(1) of such title, as added by subsection (a), and section 1787 of title 38, United States Code, as added by subsection (b)(1), shall apply with respect to hospital care and medical services provided on or after the date of the enactment of this Act.